CW00592083

Tropical forest and its environment

Tropical forest and its environment

2nd edition

K. A. Longman & J. Jeník

Copublished in the United States with
John Wiley & Sons, Inc., New York

Longman Scientific & Technical,
Longman Group UK Limited,
Longman House, Burnt Mill, Harlow,
Essex CM20 2JE, England
and Associated Companies throughout the world.

Copublished in the United States with
John Wiley & Sons, Inc., 605 Third Avenue, New York, NY 10158

First published 1974
Reprinted 1978, 1981
Second edition 1987

British Library Cataloguing in Publication Data
Longman, K. A.
Tropical forest and its environment. —
2nd ed. — (Tropical ecology series)
1. Tropical plants 2. Rain forests
I. Title II. Jeník, J. III. Series
581.909'52 QK936

ISBN 0-582-44678-3

0–470–20742–6 (USA only)

Set in Linotron 202 10/11 pt. Times

Produced by Longman Singapore Publishers (PTE) Ltd.
Printed in Singapore

Contents

Preface

Tropical forests are becoming increasingly important to everyone on Planet Earth, wherever they live. No longer can the forests be taken for granted, whether as potential farmland, an exploitable reserve of timber, a source of new biological specimens, or a place to film 'the teeming life of the jungle'. Paradoxically, their continued existence has in many tropical countries become doubtful just at the time that a clearer picture of their multiple values is emerging. Such a broader, more comprehensive view of forest land as a resource is vital, drawing upon scientific knowledge from many regions and disciplines, as well as on local experience. Indeed, an understanding of the cycling of energy, materials and genetic information, and of the life-cycles and responses of living organisms is essential to the formation of rational land-use policies in the tropics. So too is a knowledge of local conditions and history, and of real needs and constraints. Moreover, beyond the interaction of factors which control stability and limit production of natural or modified forest ecosystems, the image of a coherent global biosphere is appearing. Thus, it seems that substantial changes in the world's 2 000 million hectares of tropical forest land might have some unexpected effects on us all.

The aim of this book is to summarise available biological information on the nature of tropical forests and how they function, and to indicate some of the practical implications for anyone using or managing tropical forest land. Emphasis is placed on plant ecology and whole-plant physiology, although other botanical and zoological aspects are briefly touched on. In building up a picture of what the forests are like, questions are asked about the interactions of forest structure, composition and dynamics with differences in climatic, soil and biotic factors, and about the responses of trees to experimentally applied treatments. In considering how the natural ecosystems are likely to be affected by various management options, the key information is seen to concern the ways in which tropical forests change, and yet maintain and repair themselves, and the factors which set the limits to production and resilience.

In these circumstances, a fully revised edition of our concise book was needed, not merely to reflect the great flow of new information, but in order to emphasise for all those who wish to manage such land the relevance of the botanical knowledge about how the different tropical forest ecosystems function. The authors have been able to draw on their experience in West and East Africa, South-East Asia (including the island of Borneo), South America, the Caribbean and the Galápagos Islands, to provide a broader geographical setting and many new illustrations. The basic framework of the first edition has been retained, with its emphasis on outlining the structural and functional features of the dominant organisms of the tropical forest – its trees. Much new material is incorporated – for example, on nutrient cycling, productivity, flowering phenology and seed dormancy – and there is expanded treatment of biotic interactions within the forest.

Several entirely new sections are included on ecosystem functioning, and a whole chapter is devoted to a wide-ranging discussion of the principles and problems underlying the management of tropical forest land. The conclusions are that practical steps are possible towards sustained yield rather than over-exploitation, and towards improved and diversified value instead of declining yields and benefits. However, only where the science is really seen to help solve problems and stimulate development will it seem relevant to conserve substantial areas of what may still prove to be the most valuable natural assemblages of plants and wildlife left on Earth.

This wider coverage should appeal to planners as well as research scientists, managers as well as botanists, agriculturalists as well as conservationists, students as well as teachers. Not only should the book continue to be useful for courses in plant ecology, whole-plant physiology, forestry and natural resources management, but it may also serve to bring together the needs of forestry and agriculture, of the small-holder and the large enterprise, and of future generations as well as those of today. Only the briefest indication can be given of many relevant parts of this very large field, but an attempt has been made to provide appropriate up-to-date sources for the reader from the scattered literature, while retaining some of the classical references.

The late Mr J. B. Hall, to whose memory this second edition is dedicated, kindly provided many identifications, ecological notes and helpful corrections. Our thanks also go to Mr A. A. Enti for all his field assistance, as well as for the Asante proverb in section 7.5; to Mr J. S. Adomako for his invaluable help with many of the physiological experiments; and to Dr J. Haager,

A strangling fig in riverine forest in the Toro Game Reserve, Uganda, showing part of the extensive aerial root system.

Dr L. Soukupová and Mrs I. M. Ritchie for their assistance. We should also like to thank Květa Jeníková and Julie Longman for their encouragement and forbearance. The Institute of Terrestrial Ecology, Edinburgh, and the Institute of Botany, Czechoslovak Academy of Sciences, Třeboň, kindly provided general support, though we should emphasise that reponsibility for accuracy and for the views expressed rests with the authors alone.

K. A. L.
J. J.

Acknowledgements

We are grateful to the author, P. A. Zahl, and National Geographic Magazine for an extract on page 5 from 'Malaysia's Giant Flowers and Insect Trapping Plants', in *National Geographic Magazine of America* Vol. 125 No. 5 May 1964 p684; A. A. Enti for the translation of the Asante proverb on page 296; and to the following authors and publishers for permission to reproduce copyright illustrations:

Academic Press for fig 3.4 and fig 5.33 from 4(a) and 5 (Vazquez-Yanes & Smith 1982); Blackwell Scientific Publications Ltd for fig 4.33 from *Journal of Ecology* **58** (Lawson et al 1970), fig 6.7 from fig 4 (Edwards 1982), fig 6.14 from fig 2 (Swaine & Hall 1983), fig 6.17 adapted from *Journal of Ecology* **43, 44**. (Jones 1956); P. Cachan for table 3.1. (1963); Cambridge University Press for fig 4.11 from fig 14.1 (Jeník 1978), and fig 6.13 from fig 23.4 (Oldman 1978); The authors L. Carlos and B. Molion for fig 3.7 adapted from a figure originally published by US Airforce (AWS), Department of Commerce (Miller & Feddes 1971); Commonwealth Agricultural Bureaux for table 3.5 from *Tech. Comm.* **51**, p 24–27, "Soil under Shifting Cultivation" from (Nye and Greenland 1960); CRC Press, Florida for fig 5.28 from fig 2 (Longman 1985) copyright (c) 1985 CRC Press, Inc., Boca Raton, FL; Elsevier for fig 5.18 adapted from fig 7 (Leaky et al 1982a) and for table 3.6 adapted from table 9.9 (Golley 1983); General de Ediciones, Mexico for fig 6.15 adapted from fig 16.1 (Gomez-Pompa et al 1976); Geobotanischen Instituten, and E. T. H. Stifung, Rubell for a classification with minor alterations *Berichte Geobot. Forsch. Inst. Rubel*, **37** (Ellenberg & Mueller-Dombois 1967); International Institute of Tropical Agriculture, Nigeria for table 7.5 adapted from table 5 (IITA Research Highlights 1979); International Union for Conservation of Nature, Switzerland for fig 4.34 from figs 2.1, 2.2, 2.3, (Collins and Morris 1985); Junk, Switzerland, for figs 4.32, 5.20, 6.22, from figs 4.3, 5.8, 2.4 (Hall & Swaine 1981); Institute of Terrestrial Ecology, Penicuik and R. R. B. Leakey for fig 7.17 (Leakey & Grison 1985); The Malayan Forester for table 4.1 adapted from *Malayan Forester* **27**

(Robbins & Wyatt Smith 1964); National Remote Sensing Centre, Farnborough for fig 7.11; North Holland Publishing Company for fig 3.6 and 3.9 adapted from Schulz (1960); Martinus Nijhoff for table 6.3 from table 2 (Nye 1961); M. Murray and Format Photographers Ltd for fig 7.3; Oxford University Press for table 2.1 adapted from table 9.2 (Whittemore 1975), copyright (c) 1975 Oxford University Press; Springer Verlag, for table 6.1 from table 9.7 (Medina & Klinge 1983); Society for Experimental Biology Symposia for table 5.1 adapted from table 1 p 479 of *Symp. Soc. Exp. Biol* **23** (Longman 1969); Walter de Gruyter & Co., Berlin for fig 5.16 from abb 2 (Fink 1982)

1 Some misconceptions

This book is about tropical forests: what they are like, how they function, where they occur and why they matter. There is not just one kind of tropical forest, and yet most of them share this common feature: if the trees are removed, the land that is left behaves quite differently. It does not matter whether one is considering a mangrove forest (see Fig. 6.21), a lowland rain forest (see Fig. 2.2) or a montane rain forest (see the upper part of Fig. 1.1), they are all remarkably complex systems with a continual flow of energy, materials and genetic information passing through them. When such an area is cleared for farming

Fig. 1.1 Montane rain forest, partly cleared for agriculture, in the valley of the Río Borja in Ecuador, and the flood plain of the river. Note an epiphytic bromeliad (*Tillandsia* sp.) on the tree at the right.

(middle part of picture), much of this active cycling is abruptly modified, while slower changes will occur (lower portion) as the river gradually alters its course.

What happens to tropical forest soils in natural gaps and man-made clearings is a question which impinges directly upon the future of the plants and wildlife, and of everyone using the land, however they do so. On a regional scale, too, the presence or absence of trees can have far-reaching effects on hydrology and soils, and thus on agriculture and rural development. Moreover, recent evidence suggests that substantial alterations in forest cover may have implications for global climate that would influence the tropics and non-tropics alike.

In addition, the tropical forests probably contain nearly half of all the world's animal and plant species, including many genera and families unknown in the rest of the world. Many examples of plant structures and animal behaviour exist there which do not occur in the temperate zone, as well as specialised ecological and physiological inter-relationships between plants, animals, micro-organisms and their environment. These forests also yield a multitude of useful products, ranging from saw-logs and firewood to medicines and animal forage, which play substantial roles both in local economies and world trade. Thus these remarkable assemblages of living organisms with their habitats are important to administrators as well as scientists, farmers as well as foresters, small-holders as well as plantation growers.

On the face of it, therefore, one would expect such obviously valuable natural resources to be highly prized, and to be intentionally modified only on a small scale, except where the effects of more profound disturbance had been thoroughly tested. Quite the reverse is usually true, however, with well-tried methods rapidly giving way to new land-uses and establishment systems. Large, uniform plantations often replace the older irregular mixtures of forest and fruit trees with several farm crops. Some of the modern techniques used even involve cutting down all the larger trees and then bulldozing the whole site level, so it seems clear that scant regard is often paid to the continuing value of either the biotic or the soil resources. In essence, this paradox originates because the various interested parties disagree on how tropical forests ought to be regarded (see Fig. 1.4).

These widely differing viewpoints on environmental science and management arise partly because societies, and also forests, vary a great deal. They may also reflect the interests and expertise of the individual in question. However, there are also very many basic misconceptions about tropical forests, which colour the assumptions with which people approach them. Call a piece of land 'bush', for instance, a term which is widely employed to describe both forest and many other types of woody cover, and one is already implying that the trees are a nuisance, cluttering up a site one wishes to use in a different way, rather than being a valuable entity.

Many kinds of terrestrial and aquatic habitats have been studied for decades, and in some cases centuries, particularly in temperate regions. By combining empirical observation, theoretical work and experimental testing of hypotheses, much knowledge has been pieced together in botany and zoology, and in agriculture, forestry and fisheries. Unfortunately, there have been many attempts in the past to apply such temperate-zone assumptions and techniques directly to tropical situations, neither recognising that fundamental differences existed, nor that substantial local variability occurs. As a result, a fair amount of unsound advice has been given, and serious tropical problems have been compounded rather than solved (Fosberg 1973). In contrast, other visiting 'experts' have shown a surprising lack of interest in practical applications. One aim of this book is therefore to explore the diversity and complexity of tropical forest ecosystems, while a second is to try and narrow the communication gap between the scientists and the managers concerning the best ways of utilising them.

In the tropics, ecological investigations of grasslands and savannas have often preceded those of forests, partly because of easier access. Even in a world increasingly criss-crossed by roads and by air and sea lanes, many forests remained for a while relatively undisturbed and little known to science. In time, however, man's enlarging populations and technological capacities have led to more opportunities for study as well as to more deforestation. Similarly, the classic, standard reference book of Richards (1952) stood virtually alone for years, but since about 1970 there has been a veritable explosion of new information: papers, books, symposium proceedings, workshops, surveys, projects and programmes; on local, national, regional and international bases. Meanwhile, the relentless decline in the area of tropical forest has become more rapid, greatly exceeding natural regeneration plus replanting.

Since there is little evidence for the long-term usefulness of much of the cleared land, many searching questions remain to be answered. Are the changes which occur irreversible? Are the original tropical forests rich and productive, or not? What are the chances of changing views, because of discoveries in ecology and physiology, as well as the development of new techniques ranging from satellite survey to sophisticated biochemical analyses? What can be learnt from the older experience of centuries of farming within tropical forests? Are there unexpected links and new theories which can overstep present-day prejudices in order to forge new combinations of trees and food crops for sustained production?

Visitors from temperate latitudes are often struck by the great contrasts between tropical forests and the woodlands they have been used to. So many different kinds of trees of all sizes seem to be mixed with climbing plants to form a tangled mass of luxuriant vegetation. The frequency of aerial roots and epiphytes, combined with the sounds of strange animals, add to the unfamiliarity. As a result, innumerable travel books and articles have been written about the teeming life in the dark, equatorial jungle that is supposed to be always hot and wet.

Unfortunately the picture which these authors have painted, which has been copied into geography books and taught to children in many countries, has been for the most part misleading. As Richards (1952) has written,

> *tropical vegetation has a fatal tendency to produce rhetorical exuberance in those who describe it. Few writers on the rain forest seem able to resist the temptation of the 'purple passage',*

Fig. 1.2 Tangle of vines and other pioneers engulfing trees along the margin of a forest in S Nigeria – about 1½–2 years since clearing of adjacent forest.

and in the rush of superlatives they are apt to describe things they never saw or to misrepresent what was really there.

The following extract is by no means an exception (Zahl 1964):

We were in a forest zone so eerie it seemed bewitched. Under and around towering trees writhed lianas as thick as a man's leg, ever straining upward in search of sunlight. Other dank vegetation seemed to clutch at us like enchanted trees in a horror film. This is the homeland of gibbons, orang-utans (the legendary 'men of the forest'), wild pigs, deer and perhaps even rhinoceros. It is a haunt of cobras, and other deadly serpents too.

It may be worth noting that 20 per cent of the words in this passage do not belong to the accepted scientific 'register'.

Another reason that these travellers' tales are often misleading is that the writers are merely observing the forest from outside, while passing along a road, railway, logging track or river. Such margins frequently are an impenetrable tangle of vegetation (Fig. 1.2); but once inside the undisturbed forest it is generally possible

Fig. 1.3 Interior of mesic rain forest in W. Africa, showing the distribution of trees and absence of a well-defined herb layer. Behind numerous smaller trees and the loops of lianes is the buttressed base of an emergent *Dialium aubrévillei*.

to walk about relatively freely (Fig. 1.3), and it is seldom a question of continuous hacking of a pathway, except, for example, in bamboo thickets (see Fig. 6.19).

Moreover, the picture of the tropical forest as being permanently very hot needs a closer look, although it may appear a perfectly reasonable statement to anyone who is exerting himself. The human body is a bad thermometer, however, and the feeling of discomfort is due mainly to high humidity and low wind speed near the forest floor; indeed it is usually hotter outside. Nor is it correct to imagine that, in every tropical forest, water is continuously dripping from the leaves into a saturated soil. One may in fact often walk dry-shod shortly after rain, except in low-lying ground and for instance on some of the soils in the Andes. Similarly, to class all tropical forests as 'dark' is to ignore regional differences as well as the variations due to species, occurrence of natural gaps, and other factors which combine to govern the intricate patterns of light and shade.

Indeed, the idea of an unvarying climate is a false one, for there are always diurnal fluctuations, and also spatial differences between the upper parts of the canopy, the forest floor and the soil beneath, such that the leaves, trunk and roots of a particular tree experience quite different environments. Seasonal changes in climate may be slight or irregular in some tropical countries, but in many others they are considerable and predictable. For example, the tropical forest in the hills near Freetown, Sierra Leone, less than 8½° from the Equator, is subject to a pronounced dry season, with five months averaging only about 25 mm rainfall. During this period a dry, dusty 'Harmattan' wind from the Sahara desert may blow intermittently, sharply reducing the relative humidity for a few days or weeks, and also altering the temperature and light intensity. By contrast, in the rainy season more than 4 000 mm of rain falls, with tropical thunderstorms frequently causing disturbance to the forest. Because of the cloudy conditions, the light intensity and temperature are distinctly lower while the day-length is about an hour longer than at the height of the dry season.

It is commonly supposed that tropical trees grow continuously throughout the year, but even where the climate is much less variable than in the example just quoted, rather few do so after the seedling stage. In numerous species the expansion of new leaves and the elongation of stems is confined to periods of a few weeks, and the buds are dormant for much of the year. Growth of the cambium may sometimes be intermittent, and this has also been reported of some root growth. Nor indeed is the penetration of roots invariably shallow, as often stated. In addition, processes such as leaf-fall, flowering and seed germination generally show periodicity rather than occurring at all times.

Temperate zone misperceptions about the tropics may well lie at the root of many questionable views, both in science and management. For example, it has recently been suggested that the birthplace of ecological concepts may not have been Europe, as generally imagined, but equatorial America. Alexander von Humboldt travelled there as the nineteenth century began, and described the 'harmony in Nature' he could see in the interplay of plants and animals with the particular conditions which surround them (E. L. Jordan 1981). Nor did he content himself with theory, but frequently advocated practical solutions to problems caused by over-exploitation of useful natural resources. Thirty years later, Charles Darwin and Alfred Wallace were both to receive powerful formative stimuli in these same New World tropical forests. Perhaps it was no accident that the theories of evolution and ecology, two of the most valuable unifying concepts in biology, may have had a tropical origin. The first ecological course was also prepared and the first textbook written from tropical experience by Warming (1895), who coined the terms 'plant communities' and 'ecological factor' (Goodland 1975).

If scientists must be prepared to see some of their previous ideas turned upside-down, the fundamental paradoxes that underlie many of the changing land-use patterns in the humid tropics are unlikely to be resolved without a broader and more open viewpoint also being adopted by all the users of tropical forest land. As population pressures increase, a dozen different people – from local farmers to business investors, and from hydrologists to economists – may regard the same piece of forest in highly contrasting ways (Fig. 1.4). If sensible decisions are to be arrived at, certain basic pre-requisites will usually be needed. The first is reliable scientific information about the functioning of natural and managed forests and replacement vegetation. A second is the 'translation' of the science into language which non-specialists can appreciate. Another is the collating of local experience in utilising the forest over long periods of time. Other important steps include the carrying out of multidisciplinary research in the region, including innovative experiments, theoretical modelling and practical trials; and discussion between the local population and the authorities about priorities.

Clearly, with so many divergent viewpoints, some clashes of interest and failures of communication are inevitable. A number of preconceptions will probably have to be relinquished, both by visiting consultants and by those who live and work in the tropics,

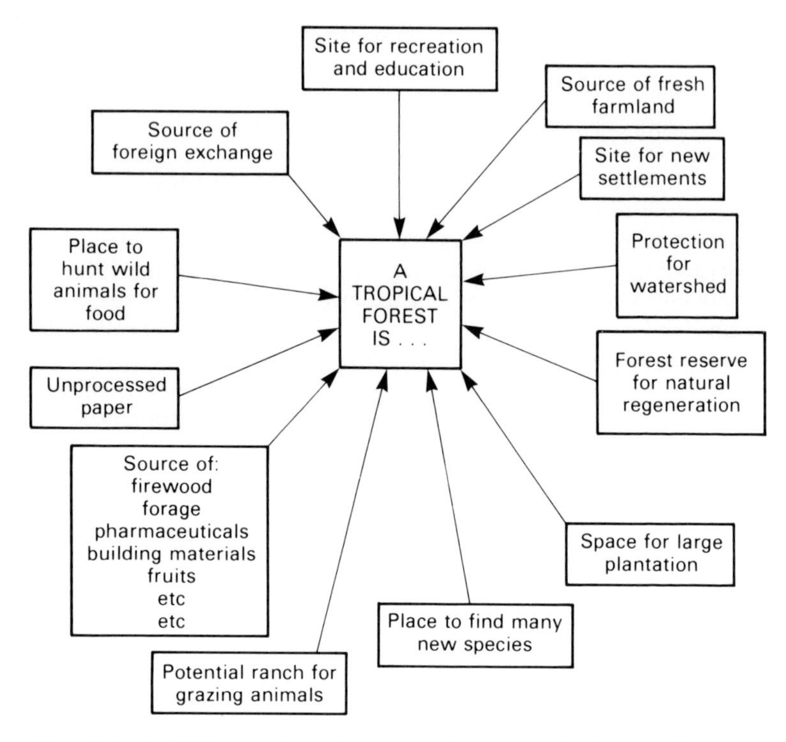

Fig. 1.4 Widely varying views on what a tropical forest is.
QUESTIONS
- Which tropical forest land-uses are incompatible with each other?
- Which allow management on a sustained yield basis?

if the major natural resource of the world's remaining tropical forests is to be understood, wisely used and conserved for future generations. Yet these are vital questions that affect the welfare, livelihood and indeed lives of several million inhabitants of many different countries. Perhaps nowhere else are so many people dependent on biological research that is still to be done, on competent assessment of the applicability of existing knowledge, and on wise decision-making and implementation by communities and governments. The problems are certainly very great indeed, but on the other hand it is evident that a world which currently spends around a million dollars a minute on armaments (Sivard 1983) could quite easily provide the resources to tackle matters that will otherwise become major sources of conflict.

2 Forest and environment interacting

The term **forest** denotes a formation dominated by trees. However, it is surprisingly difficult to define a 'tree' in the tropics, where there are numerous tree-like woody life-forms, such as tree-ferns, cycads, palms, screw-pines, bamboos and woody climbers, as well as various gymnospermous and dicotyledonous trees (for details see Hallé *et al.* 1978). The definition of tropical forests also needs to cover a wide range of terrestrial and semi-terrestrial habitats in the intertropical zone. Many local terms such as *miombo* (East Africa), *caatinga* (South America), jungle (India) have been adopted in order to express the variety of these tropical forest biomes.

As in temperate regions, three different concepts appear to be implied by the word 'forest': (1) a stand of trees or tree-like plants; (2) a whole community of plants and animals that is dominated by trees; and (3) the entire functioning system of a tree-dominated community, including its abiotic substrate and atmospheric surroundings. Definition 1 fits the viewpoint of many foresters, and indeed the demands of those concerned only with timber exploitation. The second reflects the existence of multiple interconnections between the numerous populations of plants, animals and micro-organisms associated with the trees in the forest community, and will be adopted in the following chapters. The third definition includes the participation of the environment as well as the biotic components. For the sake of clarity, therefore, explicit reference will be made to 'forest ecosystems' whenever the whole functioning entity is being discussed.

The term **environment** has different meanings, too. It may be used in the narrow sense of the various micro-environments within the space occupied by the forest, or it may broadly cover the abiotic mass and energy surroundings and underlying a particular forest site, thus approaching the meaning of the 'habitat'. The assumption will be made that both environment and habitat can be used at different levels of forest organisation, and can refer to a wide range of factors affecting the growth and development of the forest community, and of individual trees in it.

Within the variety of tropical biomes characterised by large woody plants it is often puzzling to find formations with a discontinuous canopy and grassy undergrowth. Such an 'open forest' (in French – *une forêt claire*) may occur, for example, in the transition zone with tropical savannas or with tropical marshlands, and in situations of highly seasonal water supply and/or waterlogging. The predominance of grasses and sedges in the field-layer suggests a quite different biome, and open forests are rightly classed with the grasslands. This book deals with closed forest (in French – *la forêt dense*) or, to make the distinction clearer still, with closed-canopy forest.

Fig. 2.1 Remnant of a sub-alpine forest, surrounded by man-induced *páramos*, in Sangay National Park, Ecuador. (Altitude about 3 500 m.)

Though the height of the forest structure varies a great deal due to local habitat factors, stages of stand maturity, etc., attention will be concentrated on the high forest, a term used for forest formations lying in the moderate, mesic part of the total moisture gradient encountered in the humid and seasonally wet tropics. The forest communities at the extreme edges of the gradients, for example, swamp forests, mangrove woodlands, *caatingas*, elfin forest, sub-alpine forests (Fig. 2.1), savanna-woodlands, thornwoods, etc., will be mentioned only in order to place into context the information concerning the main body of tropical forests. These are also frequently called 'the rain forests' (in the broad sense), though the same term may be used more narrowly to mean evergreen forests with little seasonality. To reduce confusion, the latter will be referred to as **rain forests**, while those with a somewhat more pronounced fluctuation of rainfall will be called **evergreen seasonal forests**. (For further details on classification see section 6.5.)

2.1 The forest ecosystem

Never in Man's history has the forest been seen merely as a collection of plants. From the Stone Age to the nineteenth century and in remote parts of the tropical world to this day, people perceive the whole of the forest's nature: its smells, sounds, moist atmosphere, dim light and humus-rich soil. Even some of its internal relationships may be noticed, like the links between certain trees and fungi or animals, and associations of particular species with certain types of soil, water or landscape. Following Tansley (1935) and the wider use of the term stimulated by Odum (1953), this network of interactions between living organisms and physical, chemical, biological and social environments is known as the **ecosystem**.

In its simplest case, an ecosystem can be represented by a single organism interacting with its environment. At its most complex, one may consider ecosystems containing millions of interacting individuals, thousands of species and large numbers of niches in one particular area (see section 4.5). This is the case with the tropical forest, one of the most diversified ecosystems found on the Earth, perhaps matched in complexity only by the underwater life of some coral reefs.

This complexity and diversity of tropical forest ecosystems is broadly accepted today. Certain of the many interactions can be readily noticed even by a casual observer, from the structures, spatial patterns and diurnal (and sometimes seasonal) dynamics of the forest. Thus it may be observed that shade-bearing plants germinate under the canopy of emergent trees, woody climbers reach the upper layer by utilising trees as scaffolding, stranglers surround the trunks of host trees, epiphyllous liverworts overgrow leaf blades, orchids grow in the crown humus, ants feed from floral or extrafloral glands, insects pollinate flowers, birds disseminate seeds, rodents feed on fruits, leopards prey on smaller mammals, and so on.

However, the pathways of life-supporting energy, flows of

materials and transfers of information cannot be so easily observed, and even detailed investigations tend to refer only to small portions of a tropical forest ecosystem. This is partly because several special obstacles hamper the ecologist and forester in pursuing their goals:

(a) The large spatial scale of tropical forests, including their partially obscured underground portions;
(b) The long time-scale of their evolution and long life-span of the dominant life-forms, mainly trees;
(c) The high diversity of organisms occurring in the forest flora and fauna.

When examining a grassland, one can obtain representative samples by observing plots a few decimetres high and several metres square. By contrast, an adequate sample plot in a rain forest would be measured in tens of metres for height and hectares for area. Therefore, in trying to comprehend either the structure or the environment one is placed at a disadvantage by the scale of the ecosystem. The forest extends around in all directions, but defies one to perceive more than a small fraction of the whole. High observation towers need to be built (Fig. 2.2) or aerial walkways such as those described by Mitchell (1982). Numerous soil pits require digging, in ground that may be very

Fig. 2.2 The euphotic zone of a rain forest, seen from 30 m up on a tower at Pasoh Forest Reserve, W. Malaysia.

hard, and root systems need to be followed down to depths of as much as 2–5 m. The many tree species present make it even harder to know whether there are local variations in structure or conditions. Thus it is not possible to make proper observations and measurements, still less to grasp the nature of a tropical forest ecosystem, from a single point on its floor.

Moreover, the long life-span of the dominant woody plants, and uncertainties about their age, cause serious problems in tropical forest investigations. Even more difficulties follow from the very lengthy periods in which the gene pools that are incorporated into the forest community develop. Thus the changing dynamics of a particular experimental or observational plot cannot be recorded by a single generation of scientists, and well established, long-term projects are needed, such as that on Barro Colorado Island in the Panama Canal (Hubbell and Foster 1983). The constraints caused by diversity of species, structure and functions can also be partially overcome by the more sophisticated types of theoretical modelling, which may allow satisfactory prediction of overall growth rates and productivity, for example, or of the effects of likely limiting factors on the survival of seedlings colonising a gap.

2.2 Geographic and climatic factors

Generally speaking, tropical forests occur in the broad intertropical zone between 23°27' North and 23°27' South. This zone, abbreviated to 'the tropics', contains about 40 per cent of the Earth's land surface. Due to the specific distribution of continents and oceans, and the circulation of air masses and sea currents, there is great climatic variation in the tropics. The geographic equator is less important than the climatic equator – the line of maximum uniformity of humidity and temperature. Seasonality in the tropics is governed by the annual march of the tropical convergence zone, which follows the position of the sun in the zenith with a time lag. Seasonality increases with distance from the climatic equator, although displacement of fronts may carry tropical air masses north (India, Mexico, Florida) or south (East Africa, Madagascar), and thus extend tropical climatic conditions beyond the geographic tropics.

At low latitudes the climate is governed by a potentially high level of incident solar radiation, due to the shorter path of the rays through the atmosphere. Köppen (1930) defined tropical lowland climates as having monthly averages of air temperature above 18 °C. Diurnal variation of temperature is normally greater than the seasonal differences, and the temperature never drops below zero. Still more important is the variation in rainfall and air humidity which forms the basis of the widely used classification of Köppen climatic types:

A_f – permanently wet rain forest: all months have sufficient precipitation;
A_m – seasonally humid or subhumid, evergreen rain forest, with a few months having arid characteristics;
A_w – dry period in the winter of the corresponding hemisphere – subhumid or xeromorphic forests, savanna-woodlands, and savannas.

Critchfield (1966) classified five types of tropical climate: rainy tropics, monsoon tropics, wet-and-dry tropics, tropical semi-arid and tropical arid climates. Tropical forests occur in the first two types and can be considered as the climax or 'zonal' vegetation of the rainy and monsoon tropics (see Fig. 2.5). The natural vegetation of the rest of the intertropical zone, as can be seen from an atlas, consists of savanna, semi-desert or even desert, depending on various anomalies in the distribution of rainfall. For instance, most of the Galápagos Islands, although lying on the Equator, 1 000 km out into the Pacific Ocean, receive less than 400 mm rainfall and are covered by arid scrub vegetation.

The principal regions with a rainy tropical climate: the Amazon Basin, windward coasts of Central America, Congo Basin, eastern coast of Madagascar, and much of tropical South-East Asia, have rainfall totals surpassing 2 000 or 3 000 mm which are distributed more or less equally over the year. Much of this area has the potential to be covered by tropical rain forest, with almost all trees evergreen. The regions with a monsoon tropical climate: the western coasts of India and Burma, a few parts of South-East Asia, the coastlands of West Africa, northern coast of South America, small portions of north-eastern Australia, and some of the Pacific Islands, may not differ greatly in total annual rainfall, but the year is divided into seasons of unequal precipitation, humidity and temperatures. The potential vegetation type ('climatic climax') of these regions is tropical evergreen seasonal forest, or alternatively semi-deciduous forest. In the former type some of the trees in the upper tree layer become leafless during the drier period, while in the latter type this also applies to the middle tree layer. The severity of dry spells in the tropical rain forests and evergreen forests is normally much less than that found in the wet-and-dry tropics with their savanna vegetation, but extreme conditions can occur (see section 3.3) that may be of considerable significance.

More details about the macroclimate of tropical countries are given in a comprehensive atlas by Walter and Lieth (1960–67), from which a selection of climatic diagrams is illustrated in Fig. 2.3. The mean annual air temperature in regions covered by tropical forests is often about 27 °C. Monthly means generally lie between 24 and 28 °C, so that as mentioned the seasonal range is less than the diurnal fluctuations which can sometimes amount to 8–10 °C (see section 3.2). Maximum temperatures recorded in the rainy and monsoon tropics rarely exceed 38 °C, thus lying far below those in the arid and semi-arid subtropical countries; indeed even in Central Europe the absolute maximum temperature may reach 40 °C.

Throughout the lowland tropical forest, minimum temperatures usually drop only a few degrees below 20 °C. For example, the absolute maximum air temperature recorded in Santarém (Amazon Basin) is 37.2 °C, the minimum 18.3 °C. At the foot of high mountains and in the bottoms of valleys, temperatures may exceptionally drop below 10 °C, mainly due to cold air currents at night, for example in Malaysia and Cameroon. The most important fact concerning temperature in lowland tropical forests is the absence of values below freezing point, which have such profound biological effects upon vegetation.

Ecologically, one of the most powerful factors controlling the pattern of tropical vegetation is the rainfall. As mentioned above, much of the area with tropical forests enjoys annual totals exceeding 2 000 mm. Favourable orography and exposure to trade winds, however, make some regions much wetter, and exceptionally up to 10 000 mm per year has been recorded (as at Débundscha, Cameroon). Recent studies in Brazil suggest that the converse is also true: that the forests influence the rainfall. At 17 sites throughout the Amazon Basin, the proportion of river water originating by evaporation from the South Atlantic Ocean or from the forest has been estimated, using the natural frequency of the isotopes oxygen-18 and deuterium as tracers (Salati *et al.* 1979). About 75 per cent of rainfall evaporates directly or via the trees, and provides much of the moisture for cloud formation and rain further inland. Significantly, deforestation near the coast seems to break the cycle which propagates the repeated succession of rain-storms moving rapidly westwards, thus threatening the survival of otherwise untouched tropical forest ecosystems far inland.

Contrary to the general impression, there is hardly any area in the world where rainfall is evenly distributed right through the year. Even in these equatorial regions of the western and central Amazon Basin, and in the Congo Basin and South-East Asia,

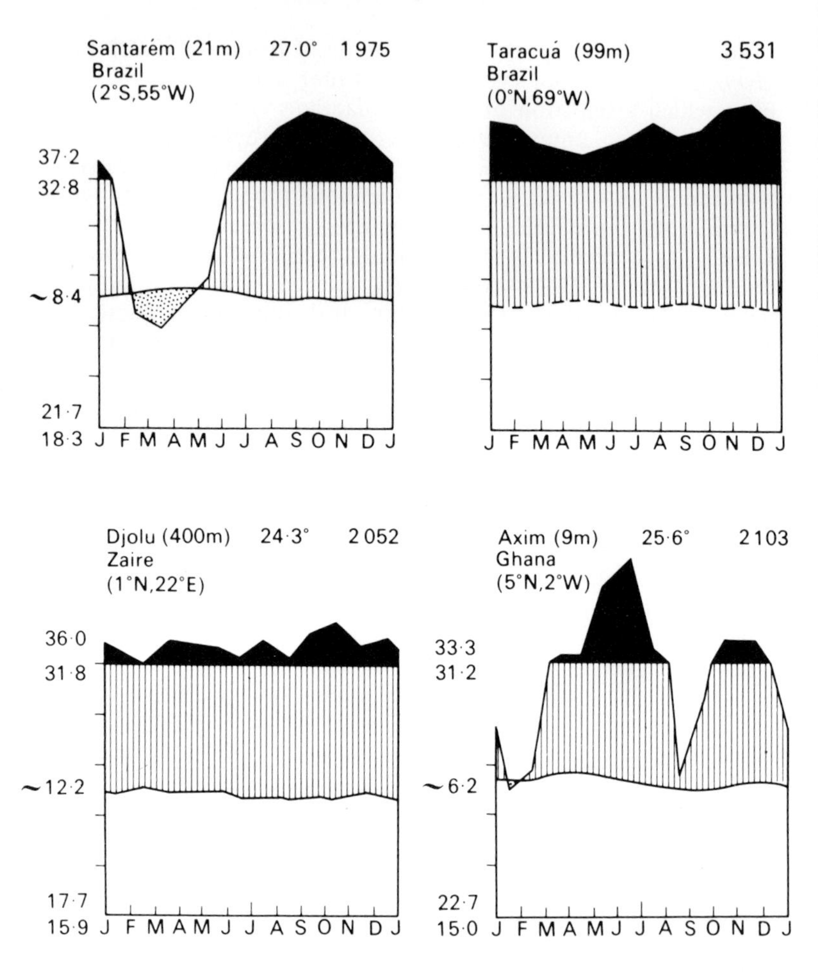

Fig. 2.3 Climatic diagrams for four contrasting sites in the rainy and monsoon tropics. The lower curve shows the mean monthly temperature (10 °C intervals on the *y* axis); the upper curve monthly rainfall totals (20 mm intervals on the *y* axis, except in the black zone representing the wet period – over 100 mm per month – when the scale is reduced by 1/10). Shaded zone = humid period; open zone = dry period.
Shown above are the altitude, mean annual temperature and annual rainfall (mm). Temperatures on the left, reading from the top, are the absolute maximum, the mean daily maximum reading for the warmest month, the mean diurnal range, the mean daily minimum reading for the coolest month, and the absolute minimum. (After Walter and Lieth 1960– 67.)

rainfall generally shows some seasonal fluctuation, or irregularities in occurrence that do not show up in monthly averages. Both the degree of dryness and the duration of the drier spell are important factors, and become more critical for the vegetation as the total rainfall decreases. Eventually tropical forests give way to savanna woodlands and ultimately to grassy savannas as the seasonal differentiation becomes more marked. Figure 2.4 shows a generalised gradient of vegetation type with respect both to rainfall amounts and rainfall seasonality.

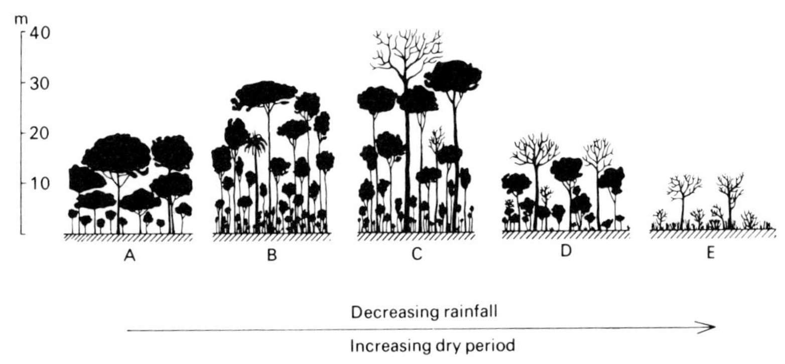

Fig. 2.4 Diagram of an ecotone or gradient in S. America from rain forests (A and B) through evergreen seasonal forest (C) and transitional forest (D) to tree savanna (E), showing that the tallest forest may not be in the wettest climate. (Based partly on Ellenberg 1959b.)

It is well known that as tropical forests merge, on their drier and more seasonal margins, into savanna woodlands, their height decreases and they have a lower standing crop. Yet it is not commonly recognised that at the rainy end of the gradient the luxuriance of the tropical forest may also decline. In Africa, for instance, the growing-stock of rain forests, and also the diameter and height of the trees growing in the wettest regions, are typically well below the figures recorded from the evergreen seasonal forests. Still more remarkable are some of the vegetation gradients in South America, where in the wettest regions of the Amazon Basin, in the area of the Río Negro, low, open stands known as Amazonian *caatingas* and Amazonian *campinas* occur (Hueck 1966). These stands – not to be confused with the dry *caatingas* found on the fringes of the Amazon Basin – are composed of short evergreen trees and shrubs with a few herbs in the undergrowth. The poorer growth in areas with frequent, heavy rainfall might be due to high rates of leaching of nutrients, or perhaps to greater cloudiness reducing light intensity and photosynthesis (see Fig. 3.7).

Any explanation and description of tropical forests would not be complete without considering past changes in climate and landscape (Livingstone and Van der Hammen 1978; Flenley 1979; Prance 1982). Mountain building and volcanic activity have clearly been important factors influencing speciation and species diversity in the tropics. The African intertropical zone was not much disturbed by mountain building and volcanic activity during the Tertiary and early Quaternary, and this could be one of the contributory reasons for its relatively smaller number of forest species. On the other hand, the South-East Asian islands, the Malay peninsula and tropical Australasia have been subjected, during the Tertiary, to periodic emergence and submergence of land areas, frequent volcanic eruptions, inundation by the sea and changing patterns of islands.

These geological modifications have affected not only the evolution of new biota, but also the patterns of dispersal and migration, and the present-day distribution of plants and animals. The profound contrasts which exist between South-East Asia and tropical Australasia, as delimited by the famous Wallace's and Weber's lines, resulted from their different geological and palaeoclimatological history. In a somewhat similar fashion, the vegetation of the Amazon Basin has no doubt been influenced by the formation of the neighbouring Andes range.

Considerable changes in the distribution and composition of tropical forests occurred during the Pleistocene, the beginning of which is estimated at only a few million years ago. Alternating humid pluvials and semi-arid interpluvials affected the extent of closed forest in relation to savannas, *caatingas* and *cerrados*. During the peaks of the glacial periods at higher latitudes, the tropical forests shrank to small refuges in the tropical lowlands (see Haffer 1982). In West Africa, for example, the area of closed forest was apparently restricted to three small refuges situated in Liberia, Western Ghana and Cameroon (Aubréville 1949). According to Ab'Sáber (1982), between 18 000 to 13 000 years BP the tropical forests of South America were represented by a series of small refuges along the foothills on both sides of the Andes, and along the northern and western coastlands of the continent. Such restrictions caused extensive depletion of the flora and fauna, but were also responsible for the origin of new taxa within each of the isolated areas. Thus, for example, the distribution of African plants, mammals and insects suggests that the West African Forest Block was separated from the one in Central Africa several times during the Pleistocene Era.

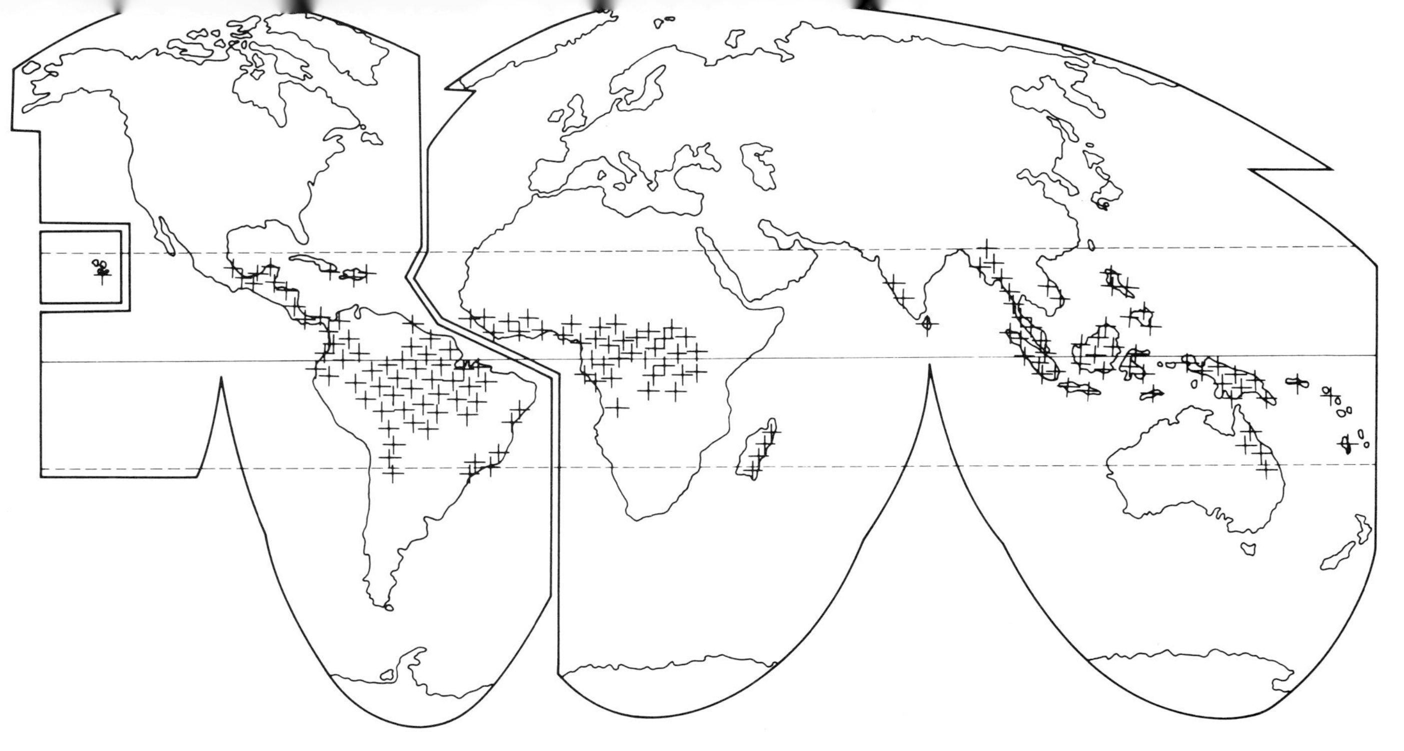

Fig. 2.5 World distribution of tropical forest land, potentially covered by tropical forests. (Inset = Hawaiian Is.)

A number of paleogeographic factors together with differences in contemporary climate, soil and human interference indicate that the world's tropical forests can be divided into four major regions (see Fig. 2.5).

1. The American tropical forest region (including parts of South America, Central America, the Galápagos Islands and the Caribbean);
2. The African tropical forest region (including the Zaire Basin, the coastlands of West Africa, the uplands of East Africa and Madagascar);
3. The Indo-Malaysian tropical forest region (including parts of India, Burma, the Malay peninsula and many of the South-East Asian islands);
4. The Australasian tropical forest region (including North-East Australia, New Guinea and the adjacent Pacific Islands).

The isolated Hawaiian Islands may represent a smaller fifth region.

2.3 Soil catenas

Next to climate, soils are the second powerful factor controlling both the distribution and composition of tropical forests. Some kinds of forest can be firmly linked with particular soil types that are true 'zonal' soils, closely reflecting the 'normal' regional climate and vegetation. However, much of the tropical forest area is occupied by soils where pedogenic processes have been affected by local features of topography, parent material or water conditions, or even by the activities of some of the biota. Soils can vary considerably over distances of as little as 10 m, this microvariability having important implications for agriculture, and also for the choice of layouts suitable for field experiments (Moorman and Kang 1978).

Many attempts to classify the diversity of tropical soils have been made. Instead of regional systems, a unified international scheme has been achieved by the authors of the *Soil Map of the World* (FAO/UNESCO 1968). Table 2.1 contains a list of soil categories important in the tropical forests, together with the approximate equivalents in other systems. The ferralsols and acrisols appear to be the main zonal soils developed under freely drained conditions. In their deeply weathered profiles, felspathic materials are decomposed slowly but completely, leaving behind simple clay minerals such as kaolinite. The mineral nutrient content is very limited, since weathering and leaching have removed nearly all bases. Only free iron and aluminium oxides,

Table 2.1 The FAO/UNESCO soil categories: approximate comparisons with other soil classifications (After Burnham 1968 and Whitmore 1975)

FAO/UNESCO (1968)	USDA (1967)	Local names, S.E. Asia (Dudal and Moorman, 1964)
Fluvisols	Fluvents; Aquents	Alluvial soils
Thionic fluvisols	Aquents	Acid sulphate soils
Regosols	Psamments; Orthents	Regosols
Vertisols	Vertisols	Grumusols
Andosols	Andepts	Andosols
Cambisols	Tropepts	Brown forest soils
Podzols	Spodosols	Podzols
Chromic luvisols	Ustalfs	Non-calcic brown soils; Red-brown earths
Gleysols	Aquepts	Low humic gley soils
Acrisols	Ultisols	Red-yellow and grey podzolic soils
Nitosols	Oxisols	Dark red and reddish-brown latosols
Ferralsols	Oxisols	Red-yellow latosols
Histosols	Histosols	Organic soils

with smaller amounts of the oxides of titanium, chromium and nickel, together with fine quartz and the kaolinite are left to form the soil. (See section 3.4 and for more details Bunting 1967 and Ahn 1970.)

However, as has been mentioned, there is in the lowland tropics great soil variability, depending on relief and to some extent on parent rock (Drosdorff 1978). Soil types with impeded, free and excessive drainage occur with accretion in the bottom and erosion at the top of the slopes. A sequence of soil profiles arranged according to topography is called a 'catena'. The lower part (alluvial complex) is represented by various types of hydromorphic soils such as gleys. These provide very different conditions for plant growth compared with the various types of ferralsols and oxisols on the gentle slopes (colluvial complex) and tops (eluvial complex). Morrison *et al.* (1948) made considerable use of the catena concept in their study of soil and vegetation in the savanna regions of East Africa, and it has subsequently been utilised in ecological analysis of tropical forests (compare Guillaumet 1967; Ahn 1970; Lawson *et al.* 1970; Markham and Babbedge 1979).

Figure 2.6 summarises three examples of catenas from African tropical forests. The first case (a) is on slightly undulating country

Fig. 2.6 Three soil and vegetation catenas from African tropical forests. (A) Weakly differentiated catena; (B) moderately developed catena, with swamp forest on the alluvial complex; (C) strongly expressed catena, with swamp forest at the bottom and drier forest at the top.

in which the upper and the middle complex of the catena remain weakly differentiated; here the soil of the gently inclined slopes with free drainage does not differ markedly from that of the crests. Consequently, a more or less homogeneous mesic forest covers most of the area, only the bottoms of valleys being wet and temporarily flooded, thus creating conditions for a tropical alluvial forest. In the second case (b), the lower part of the catena is permanently waterlogged, and is occupied by tropical swamp forest with frequent palms, with the remainder similar to (a). In the last case (c), the three complexes of the catena are most clearly differentiated: on the top there is a hardened laterite crust or bauxite layer with a drier, more open forest, even containing patches of grassland; on the slopes there is mesic forest; while in the bottom of the valley permanent swamp forest occurs.

Along the big rivers like the Amazon or Rio Negro in South America, great inland flood-plains spread over hundreds of square kilometres. The boundary between the seasonally flooded and permanently swampy part on the one hand, and the mesic well-drained area on the other, is of key importance to the local inhabitants, who make clear distinctions between them on practical and economic grounds. Thus *tierra firme* with its mesic rain forest is clearly differentiated from *Várzea*-Forest on the temporarily inundated flood-plains of the so-called white rivers and from *Igapó*-Forest in the basins of the black rivers which

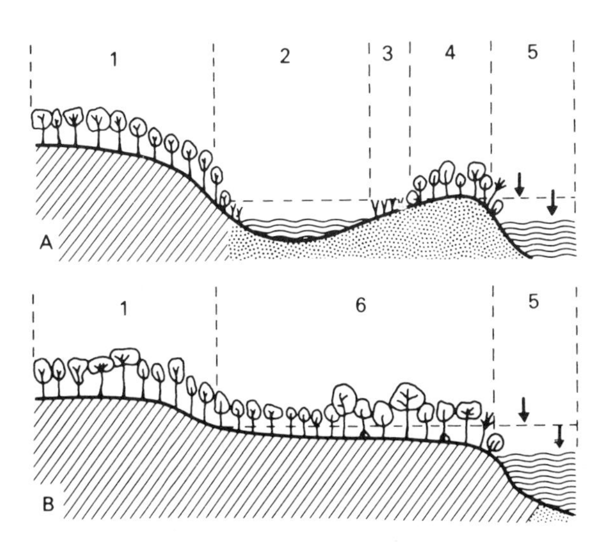

Fig. 2.7 Distribution of various kinds of S. American tropical forest, through (A) a white, and (B) a black water river valley. 1 = tropical lowland rain forest; 2 = *Várzea* lake; 3 = swamp dominated by graminoid herbs; 4 = tropical alluvial forest, here called *Várzea* forest; 5 = river bed; 6 = tropical peat forest, known as *Igapó* forest. Tips of arrows show the normal fluctuation in water level.

drain large areas of permanently wet peat forests. It has been estimated for instance that more than 80 per cent of the yield of timber from Amazonian forests has come from *Várzea* along the river banks (Volatron 1976). Figure 2.7 illustrates two catenas from the Amazon Basin.

An example of regional variation of soils, influenced by parent material and rainfall totals, is the classification by Brammer (1962) of three main soil climaxes under tropical forests in Ghana. Over intermediate to moderately acidic rocks with less than 1 500 mm rainfall, forest ochrosols (actisols) occur; they are red, red-brown and yellow-brown soils with a moderately acid to neutral soil reaction (pH 6–7). Forest oxysols (ferralsols) are a dominant soil climax where annual precipitation exceeds 1 500 mm; they develop on similar rocks to the preceding type but are much paler in colour; their reaction is very highly acid to very acid (pH 4–5). Finally, there are some forest rubrisols in West Africa; they develop over basic rocks, contain montmorillonitic clay and are dark red in colour; at surface layers they are neutral (pH 7).

Close relationships between rainfall, soil and forest composition were clearly demonstrated by Ahn (1961). In the borderlands of Ghana and Ivory Coast the distribution of various types of latosols (ferralsols) and of the forest associations followed very much the distribution of rainfall. Sample soil profiles showed the greater poverty of the wetter districts in exchangeable bases, total nitrogen, total phosphorus and organic matter; and this nutrient deficiency was apparently reflected in the height of the forest (25–30 m only), simpler layering (often with only two main strata), and less well developed emergent trees (see section 4.1, and also Fig. 2.4).

Here and there within the tropical forests even typical podzols can be found, and these invariably exert a profound impact on the composition of the vegetation. In Guyana, bleached sands give rise to so-called Wallaba forest; in the island of Borneo, heath forest is associated with similar soils (Richards 1952). In some cases these extremely acid soils with pH down to 2.8 cannot support closed forest, which, especially if degraded, may be replaced by open scrub – for example, by *padang* or *kerangas* scrubland in the lowlands of Borneo. Ultimately, especially if fire occurs, only bare rock may be left at the top of the catena.

Most soils under tropical rain forests are relatively poor in nutrients, and have seldom supported continuous farming (Fig. 2.8; but see sections 3.4 and 7.2). Nutrient supply is much better in some alluvial forests and swamps, and particularly on volcanic sediments, such as can be encountered around the volcanoes in

Fig. 2.8 Tropical forest land that has been degraded by repeated farming so that maize will hardly even grow, let alone produce any crop. Experimental area at the International Institute for Tropical Agriculture, Ibadan, Nigeria (see also Fig. 7.10).

South-East Asia or adjacent to Mt Kilimanjaro in East Africa. In the Andes, nutrient-rich andosols have developed under the impact of volcanic ash. The hydrated aluminium oxide has a very high moisture retention which can also positively affect plant nutrition.

2.4 Human influence

The third major influence on the distribution and composition of tropical forests is that caused by the interactions of their biotic components (see section 6.3). And of all the living organisms affecting the forest by far the most potent is man. When looking at small-scale maps of the world's vegetation, one may gain the misleading impression that all these large areas in the warm, humid countries of the tropics are still under a continuous cover of closed forest. However, this is now far from true, the maps usually indicating the potential vegetation that might be expected in the absence of man's interference.

Fig. 2.9 Air photograph showing a part of Nigeria where the forest is unbroken except by the river.

The amount of modification and deforestation varies a great deal, however. In some heavily populated parts of South-East Asia, for instance, much of the forest land now consists of rubber or oil-palm plantations and rice paddies. Extensive areas in Africa have been converted to farm-bush or derived savanna, and in South America to grassy ranches for grazing domestic animals. Nevertheless, there are still regions in which the original tropical forest stretches as a continuous 'carpet' over the land surface, interrupted only by an occasional large river (Fig. 2.9): portions of the Amazon and Congo basins, Borneo and New Guinea are good examples. In the rest of the areas marked on Fig. 2.5, various intermediate examples of forest modifications are widespread. It is important to realise that in most of the tropics, except for South-East Asia, dense human populations in former forest land are probably of quite recent date, even within the lifetime of the older inhabitants.

Evidence from paleontology (e.g., Leakey 1964) shows that the evolution of *Homo sapiens* took place somewhere close to the margins of the tropical or subtropical forests, probably near rivers and lakes in the open savannas. Much later, man turned to the closed forest either temporarily as a hunter or food gatherer, or because he was forced by inter-tribal conflicts to live there permanently. The tropical forest was apparently therefore not a cradle but a refuge. Later, however, sophisticated societies formed within the closed tropical forest areas, for example the Mayan civilisations in America (see sections 7.2 and 7.5), well-organised kingdoms in Africa and highly developed communities in tropical Asia. Through his social systems man was able temporarily or permanently to overcome some of the adversities of life in the humid forest.

Recently, the density of populations has been increasing sharply, due to changes in agriculture and transport, as well as the control of parasitic diseases and the availability of foreign loans. These trends are causing intense problems in the land management of many Third World countries, and questions are increasingly being asked about the future. What are the most appropriate ways of producing food from tropical forest land, for instance, and is there a connection between widespread deforestation and the severe droughts and famines that have afflicted whole regions in Africa and India?

The method of 'shifting cultivation' is still the prevailing solution which farmers in many parts of the humid tropics have found for the problems posed by the soil. Small patches of land are cleared of trees by cutting and burning, and after being used for a year or two are left to recover, usually under bush-fallow. Because of pressure on the land, however, the complete conversion of primary forest into scrub, farm-bush and secondary forest is proceeding very rapidly. In countries such as Sierra Leone, for example, farm-bush had come to prevail over forest almost completely by about 1965 (Fig. 2.10). The idea of a closed rain forest being the 'potential vegetation' becomes very hypothetical in such instances. Seeds of the majority of primary forest trees are no longer available for natural regeneration, and so the floristic composition and structure of any secondary forest may be very different from that in the primary vegetation.

Most of the tropical forest ferralsols are extremely poor in plant nutrients (see sections 2.3 and 3.4). Though these soils can carry 'luxuriant' stands of high primary production, a large proportion of the store of nutrients is contained within the living plants and in the fallen logs and litter, and is recycled with great efficiency. Thus a completely cleared farm means land stripped of its usual recapturing systems, as well as its major stock of nutrients; relatively few remain in the shallow humus layer and in the roots of woody plants. The burning of tree trunks, limbs, branches and leaves makes nutrients available for agriculture, but they are easily leached from the ash in what can be the first step

A

B

Fig. 2.10 Air photographs of farm-bush near Bo, Sierra Leone, June 1962. (A) The paler X-shaped area is a swamp with *Raphia* palms at the bottom of the catena; the village and most of the current year's farms (pale patches) are on the mesic sites. The bush fallow period is only about 5 years, and no large trees remain except near the village.
(B) Closer view of a current year's farm, burnt a few months previously. Fire-resistant oil-palms are scattered across it, and are amongst the tallest trees in the surrounding farm-bush.

to impoverishment of the site. The removal of available nutrients in the crop (or as forage and firewood) and the accelerated decomposition of litter cause further losses.

The usual practice used to be to raise two or possibly three successive food crops, and then leave the land under fallow for 10 to 15 years or more while the farm-bush or young secondary forest restored fertility sufficiently for further cultivation. However, the fallow period in Fig. 2.10 has already been reduced to about 5 years, which is not long enough for this to occur (Nye and Greenland 1960). When cultivation is (perforce) repeated even more frequently, nutrient stocks fall further still and invasions of herbaceous weeds, smothering grasses or bamboos can occur. These may tolerate fire and suppress most woody vegetation, so that various types of derived savanna, grassland and scrub (Fig. 2.11) replace the forest altogether.

One exception to these changes is found with the soils in the lower parts of catenas, which contain, receive and retain more nutrients. Thus permanent rice paddies on the flood-plains and valley bottoms have enabled South-East Asian farmers to support large populations. But on the mesic and drier sites, which are by

Fig. 2.11 Extreme example of degradation of tropical forest land. Because of frequent cutting and burning, this site near Ibadan, Nigeria, carries grassy vegetation with *Opuntia* instead of tall tropical trees.

far the most common, the position remains open. Are methods available that would allow sufficient food to be grown each year to feed everyone adequately? Could a proportion of the tropical forest be retained in each area, because of the many other 'goods and services' it provides? What about all the other land-uses competing for the same sites (see Fig. 1.4)? These and other vital questions will be further explored in Chapter 7, against the background of the steadily depleting tropical forest resources of the world. Indeed, unless the ecological factor, man, places himself under reasonable control, some of the present options may soon disappear altogether (see Fig. 7.3).

3 Environmental factors

The geographical latitude, circulation of air masses, underlying bedrock and topography are the external pre-requisites or conditions for the internal environmental regime within a tropical forest ecosystem. However, all the external sources of energy, materials and information resources that are made available at particular locations are considerably transformed by the trees and the accompanying biota. A dynamic series of gradual transductions, adjustments and alterations of each type of environmental factor is inseparable from the processes of growth, reproduction and activity of the living inhabitants. This network of internal components and relationships is indicated in an approximate fashion with the help of concepts ranging from prerequisite and condition on the one hand to factor and trigger factor on the other.

As with other terrestrial ecosystems, a number of common concepts are used to describe and model the whole tropical forest system. To begin with, physical and biotic factors are separated, or (more precisely) the abiotic and biotic factors of the environment. The contents of the ecosystem are divided into inorganic and organic components, and furthermore the useful distinction between live and dead organic matter is often adopted. It is usual to separate the live inventory of ecosystems simply into plants and animals, and perhaps to recognise the vague category of micro-organisms, but alternatively one may adopt a more advanced basic classification into monera (bacteria and cyanophyta), plants, fungi and animals. A spatial division of the terrestrial ecosystem into its soil and atmospheric portions may be used, or a simple partition into underground and above-ground spheres. Taking into account the most important interfaces between biotic and abiotic components of the forest, we can distinguish the **rhizosphere** (the zone of interacting roots and soil), and the **phyllosphere** (the zone of interacting leaves and atmosphere).

Tropical forests, however, pose frequent difficulties for the adoption of such simplified concepts. Conditions may easily be trigger factors, physical and biotic factors are inseparable, live

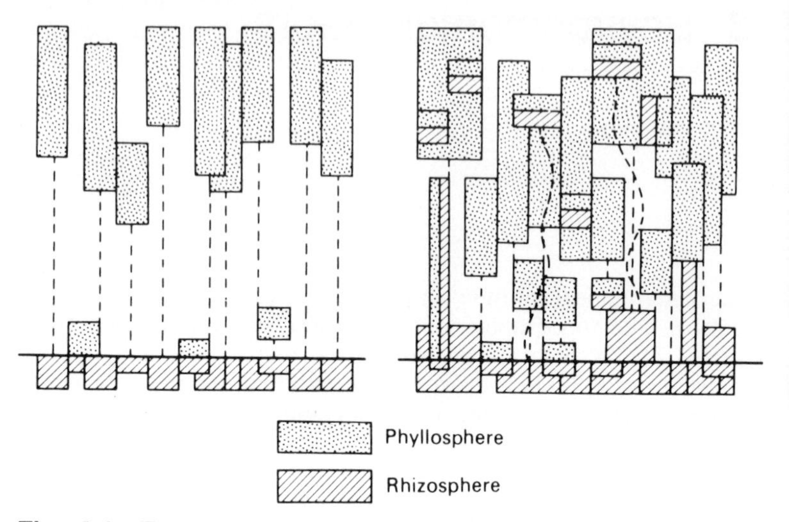

Fig. 3.1 The spatial distribution of rhizosphere and phyllosphere in *left* – a temperate beech forest; and *right* a tropical rain forest.

and dead material are not easy to separate, and the variety of biota can hardly be expressed by the handy portmanteau terms 'plant' and 'animal'. The very nature of 'soil' is rather difficult to define among so many different substrata, since soil, soil atmosphere, aerial soil pockets and atmosphere are spatially confused. Thus, in contrast to temperate forests, one cannot make a complete distinction at ground level between the rhizosphere and phyllosphere (see Fig. 3.1). Soil pockets occupied by plant roots may occur at any height above the ground: in fissures in the bark, in forks between big limbs, on top of horizontal branches, in the hollows between adjacent leaf bases. Small water tanks may appear high in the tree crown – for example, in depressions on leaves, branches and tree trunks – creating an environment for aquatic algae, vascular plants, insects and amphibians.

Spatial distribution of different micro-habitats, and a partially obscured distinction between biotic and abiotic phenomena make it difficult to select a few dominant factors and treat them in isolation. Nevertheless, the transformation of the sun's radiant energy and the consequent distribution of light and temperature are important above-ground aspects of the environment that clearly need individual treatment. The water regime and the cycling of nutrients are more closely related to the substratum of the forest, and these also deserve special attention. Disturbances to the ecosystem, caused by powerful forces such as wind and fire, are also singled out in this chapter. The effects of environmental factors on tree growth are considered in Chapter 5, and the dynamic functioning of the ecosystem as a whole in Chapter 6.

3.1 Light and shade

Almost all the energy in tropical forest ecosystems originates from solar radiation, which strikes the canopy and ground surface in a broad spectrum of rays. Less than 2 per cent of this radiant energy from the sun occurs at wavelengths longer than 4 μm, which is used as an arbitrary boundary between short- and long-wave radiation. On the other hand, there is a much higher proportion of the relatively energy-poor, long-wave part in the radiation that is released back to space by plant and ground surfaces. The energy-rich, short-wave solar radiation is the portion that is important in the photosynthesis of green plants, especially the region between 400 and 700 nm known as photosynthetically active radiation (PAR). The absorption, reflection and diffraction of radiation of various kinds within the canopy also determines the temperature, and consequently the rates of many physiological processes in plants, micro-organisms and animals that do not maintain a specific body temperature.

Visible radiation sources in general have been a powerful factor in the evolution of forest plants, with broad differences in growth form and physiological adaptations between light-demanding (heliophilous) and shade-bearing (sciophilous) species. Particular regions of the spectrum control various photomorphogenetic processes, including the responses to day-length and the germination of light-sensitive seeds (see Ch. 5). Specific parts of the short-wave radiation are 'visible' for various animals, and are responsible for the timing and periodicity of their activity. Those active in the day-time often have a different visual perception from nocturnal species. Therefore, although we may in a simplified way look upon the PAR wavebands as 'light', and the alternatives as 'shade' and 'darkness', these differences in the visual capacity of other vertebrates and invertebrates should not be forgotten.

The tropical forest markedly transforms the light striking the surface of its canopy. Ecologically important are both the quantity, that is the distribution of its energy (measured in illuminance, irradiance or the flux density of photons); and the spectral quality, or its content of the various coloured wavebands. The upper tree layer, with the fully exposed emergent trees and their associated epiphytes and animal life (section 4.1), exists in the

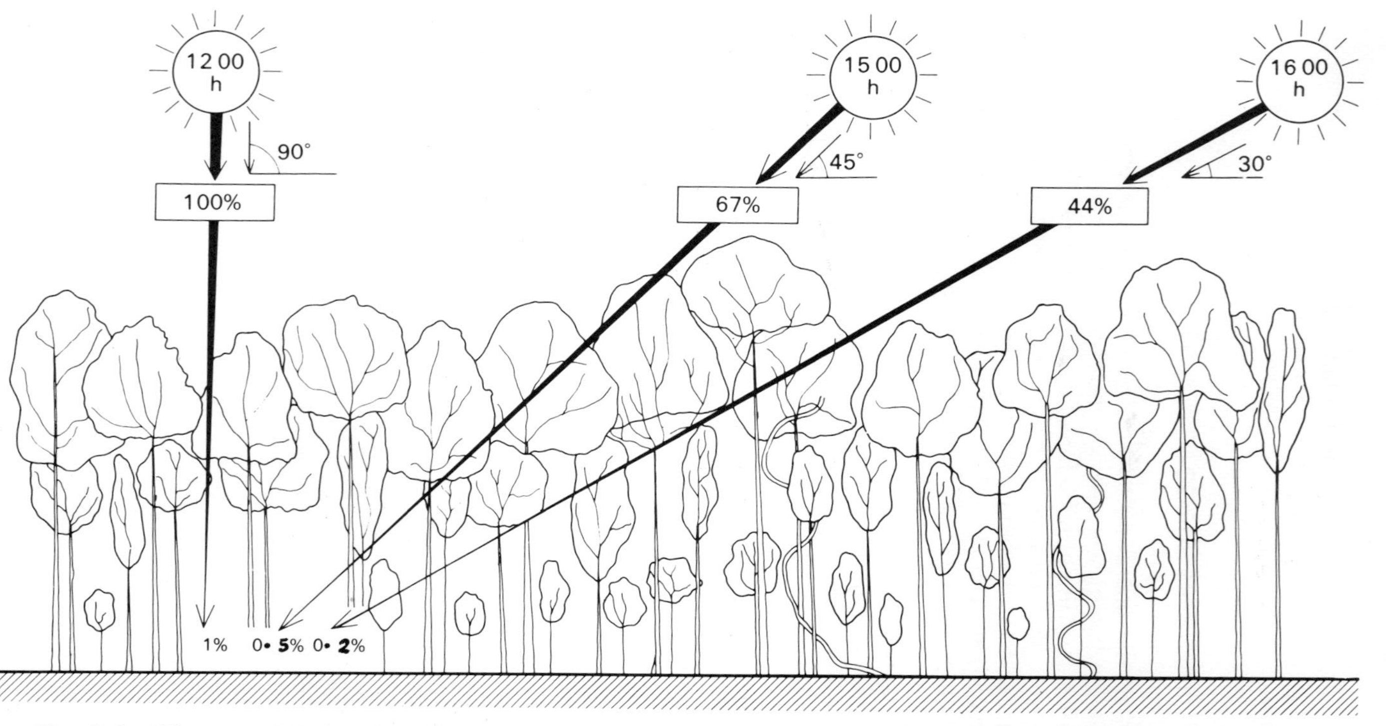

Fig. 3.2 Diagram showing the effect that the structure of tropical forest has upon the penetration of light from different angles. Changes in the relative light intensity above the canopy and at the forest floor, taking as 100% the value above the canopy with the sun overhead at noon. (Jeník and Rejmánek, unpublished material.)

euphotic layer of the forest (Richards 1983). This receives from 25–100 per cent of relative illuminance (Fig. 3.2), and often the uppermost twigs carry densely packed foliage, may produce flowers and fruits prolifically, and bear heliophilous epiphytes such as orchids, bromeliads and lichens (Fig. 3.3). The euphotic layer is the most productive part of the forest, originating much of its biomass and its diversity of animal life.

Lower down, the relative illuminance decreases markedly as the **oligophotic** layer is reached, throughout which there is usually less prolific growth of plants and less animal activity. Below the middle tree layer, the relative illuminance falls to 1–3 per cent, and in the lower tree layer epiphytes are either less numerous or are represented by sciophilous species, such as ferns, mosses and liverworts. In the oligophotic layer as a whole, quite strong competition for light takes place, and various adaptations of plants and arboreal animals are found. The minimal light supply near the ground surface is reflected in the low development here of plant biomass and in the stunted growth form of any seedlings, saplings and pigmy trees.

Some forest plants in this shaded environment exhibit a conspicuous blue-green iridescence of their leaves, as for example *Selaginella* spp. The blue colour is caused by an interference filter (Lee and Lowry 1974) that increases reflection of photosynthetically rather less active light (400–500 nm) and increases penetration of the most active wavelengths (600–680 nm). In addition, such plants may possess egg-shaped cells in the epidermis, with a convex outer surface and chloroplasts in a peculiar position at the inner end of the cell. With their interference filters these cells resemble a camera, complete with coated lens. It has also been suggested that the conspicuous red pigments, found for example in the vacuoles of the lower epidermis of *Begonia* leaves originating from such dim light habitats, may be related to a 'back-scattering' of useful red light again into the photosynthetic tissue (Lee *et al.* 1979).

Many ecologists and physiologists have estimated relative illuminance in the undergrowth to be lower still, even a fraction of 1 per cent. For example, Bünning (1947) found the relative light intensity in the interior of an Indonesian rain forest was only 0.2–0.7 per cent; Carter (1934) reported very similar data in the *Mora*-forest of Guyana, as did Cachan (1963) in a rain forest in the southern part of Ivory Coast. Yoda (1974) gave 0.3 per cent as a mean relative illuminance in an undisturbed rain forest of West Malaysia. Cachan and Duval (1963) and Cachan (1963) were among the first to record vertical gradients of illuminance in an entire profile of a tropical forest. Using a steel tower they

Fig. 3.3 View of the upper portion of the trunk of *Shorea leprosula*, covered with crustaceous lichens growing in the euphotic layer. Photographed 25–30 m up the tower at Pasoh Forest Reserve, W. Malaysia.

were able to measure the light intensity from the top of the emergent trees down to the forest floor (see also section 3.2). Their data show a profound screening effect of the closed canopy in the middle layer: while the illuminance at 46 m at the top of the forest was 100 000 lux units, it was 25 000 lux at 33 m, still above the closed middle layer, and simultaneously only 800 lux at 1 m from the forest floor, 0.8 per cent of that above the forest.

The quality of the light within the canopy and near the ground in tropical forests is affected by two components: (1) the incidence of unfiltered daylight composed of direct solar irradiation, diffuse skylight and diffuse light reflected from the clouds, and (2) filtered daylight, the spectrum of which has been transformed by the reflection, absorption and transmission processes of the foliage, twigs and stems. Leaves play a major role in the filtering of daylight, for they are effectively opaque in the range of visible wavelengths (400–700 nm), except for a small portion in the green band around 550 nm. Beyond the limits of 700 nm a dramatic increase in transparency occurs, and causes a high proportion of far-red wavelengths (FR) to penetrate deep into the forest. High representation of far-red radiation is particularly important in relation to the visible near-red band. The low R/FR ratio occurring within the forest appears to be responsible for numerous morphogenetic processes, notably as a factor maintaining the dormancy of seeds in some colonising species. Vázquez-Yanes and Smith (1982) showed experimentally that, for the germination of *Cecropia obtusifolia* and *Piper auritum* seeds lying on the surface of the soil in the forest, the R/FR ratio is much more important than the light intensity (see Fig. 5.33).

Since tropical forests are three-dimensional structures (section 4.1), their patterns of radiation and illuminance also vary horizontally. On the floor of a mature and undisturbed forest it is often possible to distinguish three phases: a **dim phase** where there is the 'normal' thickness of crowns, tree trunks, dwarf trees and seedlings; a **light phase** where there are small gaps created by fallen trees or broken limbs of emergent trees, with many seedlings germinating and sprouting; and finally, a **dark phase** where there are thickets of dying branches and living climbers which have fallen down from the upper layers without disrupting the canopy seriously. Only a few ground herbs and no tree seedlings can be found in the third phase. The size of the light phase gap affects both the intensity and the spectral composition of the extra sunlight. Surprisingly, detection of the F/FR ratio in *C. obtusifolia* is so precise that few seeds germinate unless they are situated away from the margins of gaps exceeding a certain size (Vázquez-Yanes and Orozco Segovia 1984, see Fig. 3.4).

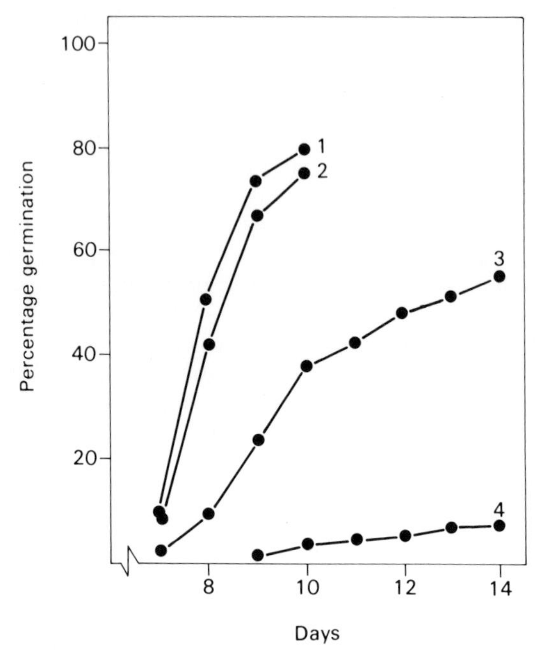

Fig. 3.4 Differential germination of light-sensitive *Cecropia obtusifolia* seeds placed in petri-dishes across a crown gap in a Mexican rain forest, caused by the fall of a 30 m *Spondias mombin* tree. Positions 1 and 2 near the centre of the gap; 3 and 4 progressively closer to the edge; 5–8 at the edge and inside the forest, where the low R/FR ratios kept all seeds dormant for the month's duration of the experiment. Total distance covered – 20 m. (After Vázquez-Yanes and Smith 1982.)

Another important source of diversity in the illuminance of the forest interior is caused by the small patches of light called 'sunflecks', in which sunlight breaks through a hole in the canopy (Fig. 3.5). Sunfleck-light is composed of direct sunlight, sunlight reflected from vegetation, diffuse skylight and diffuse skylight filtered by the vegetation (Morgan and Smith 1981). The sunflecks move as the sun changes position, and as the wind or ascending air current disturbs the leaves. Their spectrum and flux density vary considerably over their ever-changing positions, and, remarkably, the R/FR ratio at their periphery may be lower than in the ambient shade. Moreover, at a certain spot, the photosynthetically active radiation might vary over two orders of magnitude within a few seconds or minutes. Thus, the leaves must be adapted to utilising very low flux densities, but also tolerating and using the high fluxes in sunflecks from direct solar beams (Kwesiga 1985). It has been reported, for instance, that stomata in seedlings of *Shorea leprosula* and *S. maxwelliana* open within seconds of the arrival of a sunfleck (UNESCO 1978: 180).

Fig. 3.5 Interior of S E Asian forest, showing sunfleck-light and the base of a *Shorea macrophylla* tree.

The light intensity near the forest floor thus depends partly on the structure of the canopy, but also on the angle at which the sun's rays are striking its surface. The low angle of incidence in the early morning and late afternoon hours increases the path length of the light rays through the canopy (Fig. 3.2). The relative illuminance on the forest floor is considerably diminished, since the oblique rays are obstructed by more leaves, twigs, limbs and trunks, and the number of sunflecks also decreases. Leaf-movements, particularly common in leguminous trees, can increase this tendency (see section 5.4), and it can even be quite difficult to read instruments correctly within the forest during the first 2½ h of the day, and again in the last 2½ h, because the light is so poor. Figure 3.6 is an example of the daily march of illumination.

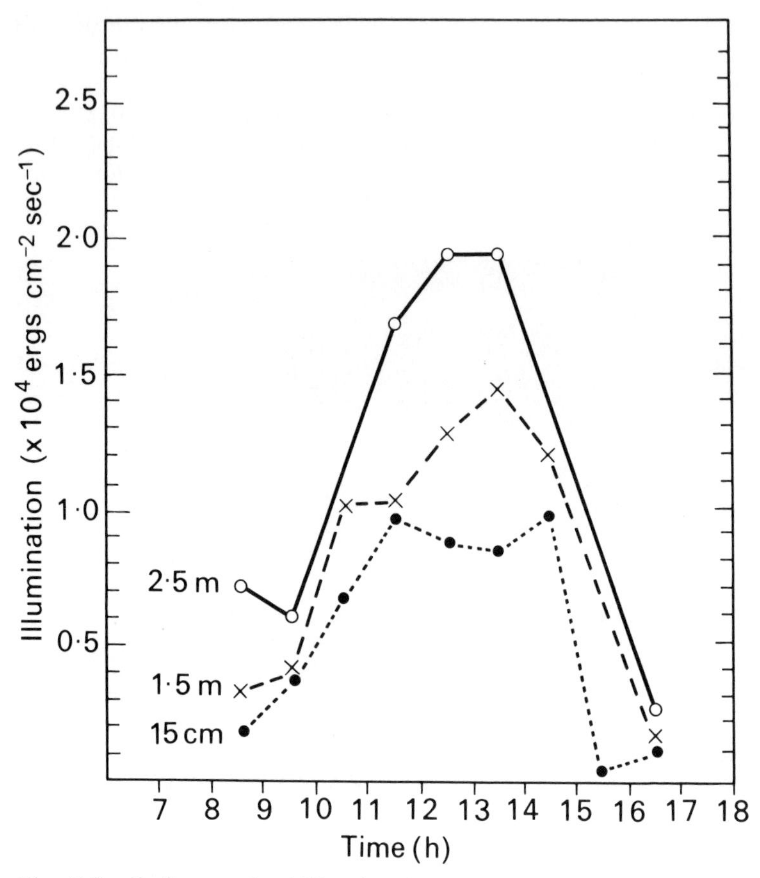

Fig. 3.6 Daily march of **illumination** in a rain forest in Surinam, at 3 levels above the ground. (After Schulz 1960.)

Light intensity and spectral composition also change seasonally for various reasons, such as the shedding of leaves (section 5.5), and the incidence of hazy periods (see Fig. 5.35) and particularly of clouds. The composite satellite pictures of average cloudiness in South America displayed in Fig. 3.7 show that there is considerable variation, even in parts of the Amazon Basin that are on the Equator. Such seasonal changes will clearly affect the photosynthesis of the forest as a whole, and cloudiness may also modify or remove the sunfleck-light component, and therefore alter both the flux densities and the R/FR ratios which are so physiologically important to the forest floor plants.

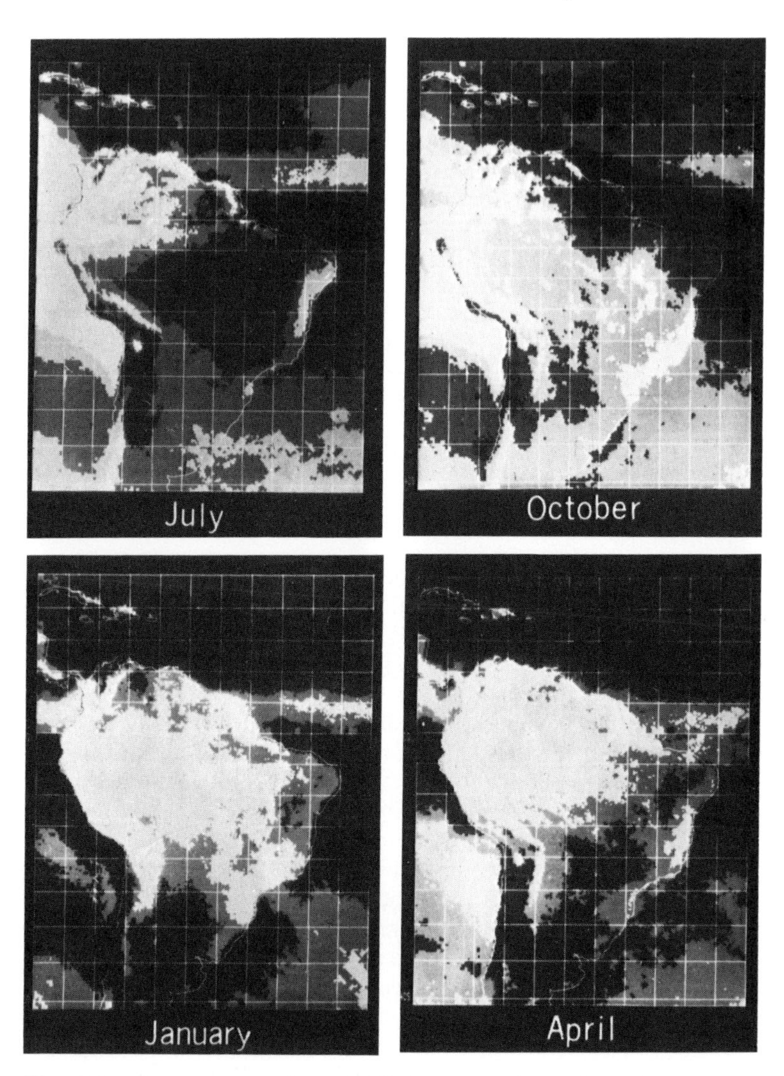

Fig. 3.7 Four composite satellite images of monthly cloud cover over S. America, averaged for the period 1967–70 at 1400 h local time. (From Miller and Feddes 1971.)

Often overlooked, the day-length or photoperiod also varies appreciably in the tropics. At 5° latitude, the variation is just over half an hour, at 10° it exceeds one hour, while at 17° there is as much as two hours difference between June and December. As shown in Chapter 5, alterations in day-length of this magnitude can result in substantially modified growth and development of tropical trees, irrespective of changes in daily flux density that may also occur. In calculating the natural day-length an allowance must be made for dawn and dusk, and for whether the particular site is near or under the canopy of the forest, since plants are sensitive to quite low intensities and to the R/FR ratio in their photoperiodic responses. On the Equator, and when the sun is overhead elsewhere in the tropics, the effective day-length is about 12¾ h for the tops of the canopy (or the field layer of a very large gap in a flat landscape with a clear sky). It would be less inside the forest or in steep terrain, and in cloudy weather, or with a species requiring above-average intensities for photoperiodic control.

3.2 Temperature variations

The monthly average air temperatures at most meteorological stations that are presented in atlases and in climate diagrams are proverbially equable (Fig. 2.3), and some writers have even concluded that temperature is a generally favourable factor which need not be considered in the tropics. However, lying behind these monthly means, which are derived from relatively few stations mostly situated in open ground, considerable diurnal, periodic and spatial variation is generally hidden. Neither the ecologically decisive maxima and minima that occur within the three-dimensional structure of the forest, nor the physiologically effective diurnal fluctuations in temperature will be shown up. One should also consider the possibility that within an equable macro-climate even small changes of around 1–2 °C might have as much effect on plants, animals and people as differences of 10 °C in highly seasonal climates (see Ch. 5).

Furthermore, low temperature stress can occur not only at 0 °C and below, but at non-freezing temperatures below 10–12 °C or even in some species below 15–20 °C. These cause *chilling injury* in plants, and may lead to physiological stress in animals. The mechanism for chilling injury encompasses several elements operating independently or simultaneously: imbalances in metabolism, accumulation of toxic compounds, and increased membrane permeability (Lyons 1973). Both the people who live there and explorers in camps within the tropical forests are well aware of the discomfort felt when night temperature falls only a few degrees lower than usual.

Superimposed upon the diurnal and seasonal fluctuations,

considerable differences in temperature occur between the various layers of the forest, and if the canopy is interrupted by natural or artificial breaks, large differences also occur horizontally. Some detailed measurements of these variations have been carried out in South America, Africa and Malaysia. In the rain forest of the Ivory Coast, Cachan and Duval (1963) used their high tower to study also the vertical temperature gradients throughout the year. In December, the diurnal range of temperature (calculated in weekly periods) was 10.8 °C at 46 m, near the top of emergent trees, but 4.4 °C at 1 m height in the undergrowth. In June, the corresponding figures were only 4.0 °C at 46 m, and 1.7 °C at 1 m height. Data for the entire profile through the forest are presented in Table 3.1. The absolute maximum range recorded at 46 m was 14.5 °C, in January; the absolute minimum range at the same level was 1.5 °C in June. Comparable figures at 1 m height were 5.9 ° and 0.7 °C respectively.

Table 3.1 Diurnal range of air temperature (calculated for a weekly period) at various heights in a closed rain forest in the south of the Ivory Coast. (After Cachan and Duval 1963.)

Height above ground (m)	December (°C)	June (°C)
46	10·8	4·0
33	10·0	3·8
26	9·9	3·4
12	6·6	2·8
6	5·2	2·2
1	4·4	1·7

Moreover, throughout the year and down the vertical profile of the forest, the important parameters of daily maximum and night minimum temperatures can vary a good deal. Since averages have prevailed in the climatological treatment of tropical environments, relevant records of extreme values of temperatures are relatively rare. Table 3.2 shows data from the same study area in the Ivory Coast, which reflect the seasonality found even in the wettest parts of the West African forests. Parallel deductions can be made concerning microclimate in tropical forests in general – for example, the remarkable similarity of minimum temperatures along the whole vertical profile.

Allee (1926) pointed out more than 60 years ago that above

the tree canopy temperatures show a close resemblance to those recorded at 1.5 m height in an extensive clearing. Taking into consideration the technical difficulties of working near the top of the forest, one can appreciate the value of supplementing such studies by comparing near-the-ground measurements within the forest and in an opening outside it. Schulz (1960) collected very full data on the temperatures in different habitats of the tropical forest in Surinam. The daily maximum in the forest undergrowth usually varied between 25 and 30 °C, and the minimum between 20 and 22 °C.

Table 3.2 Maximum and minimum air temperatures (calculated for a weekly period) at various heights in the rain forest of the Ivory Coast. (After Cachan and Duval 1963.)

Height above ground (m)	Maximum temperature (°C)		Minimum temperature (°C)	
	July	February	August	February
46	25·8	32·3	18·1	23·1
33	—	31·1	—	23·1
26	24·1	30·3	18·2	23·1
12	23·8	30·0	18·7	23·2
6	—	28·5	—	23·3
1	23·8	28·2	19·2	23·4

Complete curves comparing the diurnal march of temperatures in the forest interior and in a clearing are shown in Fig. 3.8 which is derived from an ecological study of the Ankasa Forest Reserve in south-west Ghana (Jeník and Hall, unpublished material). This was done in a 'Harmattan' period, which in West Africa occasionally brings rather extreme conditions even to the rain forests (see Fig. 5.35). In other tropical regions there can also be periods of irregular temperature change. Hueck (1966) mentions the abrupt drops in temperature in the eastern part of the Amazon Basin during June, July and August, the so-called 'friagems'; and similarly in Central America *nortes* can occur. At the foot of high mountains in Malaysia, the temperature may exceptionally fall below 10 °C even in lowland dipterocarp forest.

In more seasonal climates with their corresponding semi-deciduous forests, air temperature varies substantially during the year, and also spatially: for example, in patches of flood-plain forests and adjacent climax forests. In Costa Rica, Janzen (1976*a*) found little differences of temperature between evergreen

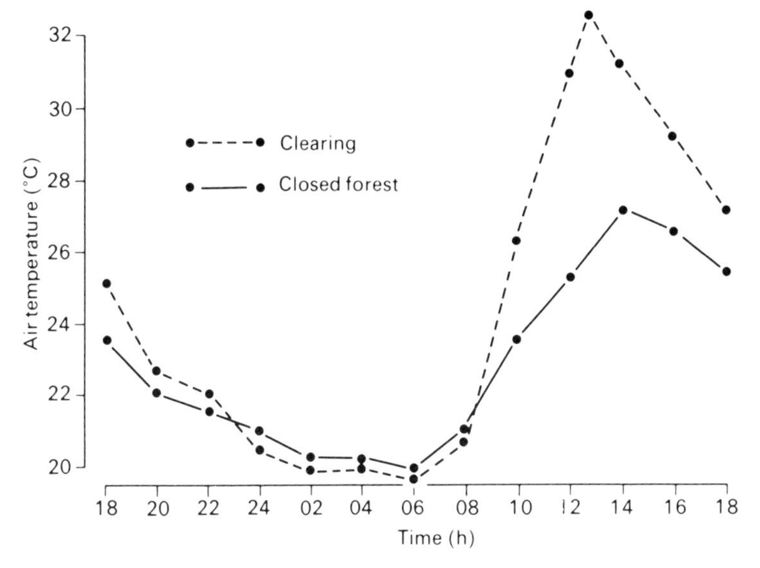

Fig. 3.8 Daily march of **air temperature** within a rain forest in S W Ghana, compared with a large clearing. Both at 10 cm above the ground.

riparian forest and nearby deciduous forest during the rainy season, but near the end of the dry season air temperatures in the former stands were as much as 5.5 °C cooler.

Soil temperatures within tropical forests in general show only minor seasonal and diurnal fluctuations. The maximum soil temperature under closed forest probably never exceeds 30 °C, in contrast to open clearings where surface layer temperatures may temporarily surpass 50 °C. Inside the forest, soil temperature seems to be controlled largely by average air temperatures, and is only slightly affected by changes in soil moisture. A comparison of the daily march of soil temperatures under closed tropical forest in a small gap and in an open clearing is given in Fig. 3.9 using data published by Schulz (1960). The same author calculated that the yearly average soil temperature was practically equal to the average air temperature in the undergrowth of the forest. At or below a depth of about 75 cm there is no diurnal fluctuation of soil temperature, and the seasonal range is very small indeed. Over a period of more than two years, Schulz recorded an absolute extreme range of temperature at this depth of just 1.5 °C, suggesting a surprising equability of environment for root growth and soil organisms here. It therefore follows that a single measurement made in the subsoil can be sufficient to

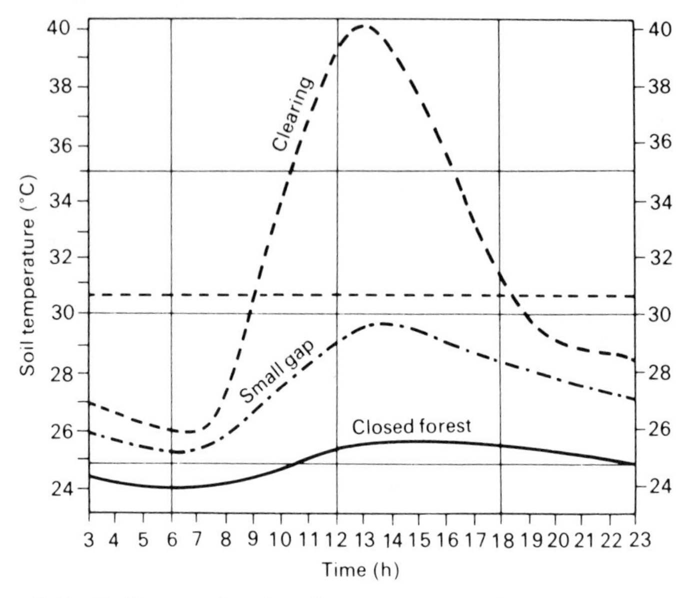

Fig. 3.9 Daily march of **soil temperature** in a rain forest in Surinam, in closed forest, a small gap and a large clearing. All measured at 2 cm depth, averaged over the dry season. The horizontal line is for 75 cm depth in the clearing. (After Schulz 1960.)

indicate the annual average soil temperature at this depth, and may even provide a satisfactory estimate of the annual average air temperature in the undergrowth of the tropical forest!

3.3 Water conditions

As explained in section 2.2, the boundaries of the area which could potentially be occupied by tropical forest can be broadly correlated with the amounts and seasonal fluctuations of rainfall. Thus water supply is justifiably considered to be a key factor in the life of the tropical forest, playing an essential role in the growth and reproduction of its component organisms and in the productivity of the whole forest. Many plant ecologists and geographers have incorporated features connected with moisture and rain into terms defining various kinds of tropical forests. Schimper's expression (1898) *der tropische Regenwald* has been widely accepted and its equivalents, such as *pluviisilva*, tropical rain forest and *la forêt ombrophile*, coined in other languages.

In many of the subdivisions of tropical forests used by Richards (1952), Whitmore (1975) and other writers, there is frequent use of words denoting atmospheric humidity and/or soil moisture: moist forest, everwet forest, wet-land forest, mist forest, cloud forest, swamp forest, freshwater forest, brackish-water forest, etc. Similarly, a range of terms denoting dryness is found.

Under the influence of these moisture-dominated descriptions, students in botany and forestry with little experience of tropical conditions tend to exaggerate the frequency of rains, saturation of air with water vapour, swampiness of soil and absence of water deficits. Most tropical forests are in fact mesic (or mesophytic) vegetation types, situated on freely-drained soils with moisture conditions quite unlike those of swamps. The variation of soil and forest along catenas (section 2.3) has to be considered in each case, though it was commonly neglected in older enumeration surveys of tropical forests.

Moreover, the temporal variation in rain and humidity plays an important part in forest life. Seasonal forests regularly experience very different conditions in the dry and rainy seasons, while ecosystems that are normally continuously moist can be strongly affected by rainless periods of a few weeks' duration. Although such events may occur only occasionally, they appear sometimes to be associated with mass flowering (section 5.9), while rarer but more prolonged droughts can lead to profound alterations in the forest (section 3.5). Diurnal fluctuations should not be overlooked, nor the widely differing micro-environments experienced amongst the tree tops and on the forest floor during day-time. The 'from-the-bottom' view can be very misleading here, as can ideas based on temperate forests, which tend to have only slight differences in atmospheric conditions within them, perhaps because they are often aerodynamically 'rougher', and of shorter stature.

Rainfall is the primary source of water in mesic tropical forests, while many kinds of montane and subalpine forests also experience other kinds of precipitation, such as drizzle, low clouds, mist and fog. Riverine and flood-plain forests and swamps in badly drained depressions do not have to rely only upon local atmospheric precipitation, since they can utilise water resources arising from neighbouring or distant humid or per-humid regions, such as mountainous areas. Tropical forest contributes strongly to the recycling of the re-evaporated moisture in certain water basins. In the Amazon Basin, several estimates show that inflowing moisture from the Atlantic Ocean accounts for only about a half of the precipitation inside the region, while the other half originates from water vapour recycled within the area (Salati *et al.* 1979). The four pictures in Fig. 3.7 showing 4-year averaged cloud patterns over South America provide some suggestion of how this long-lasting and recycling humidity spreads across the continent.

As in other types of vegetation, the amount of precipitation falling on tropical forests can be partitioned according to the relationship:

$$P = I + SF + TF$$

where P is precipitation, I is water intercepted by the vegetation and evaporated directly, SF is intercepted water that subsequently reaches the ground by stem flow, and TF is water that either falls directly on the ground through gaps in the canopy, or drips from the foliage (Doley 1981). Table 3.3 contains a few examples of the partitioning of precipitation in some tropical and subtropical forests and woodlands.

The structure of the tree canopy and even the pattern of the lower layers is decisive for the proportion of evaporated, intercepted and through-fall water. The precise mechanisms of this partitioning are more complicated than in temperate forests, where epiphytes, climbers and aerial roots play little or no part. The density of trees, limbs, branches and leaves is only the prerequisite for processes that are further affected by the orientation of branches and limbs, by the morphology of leaves and leaf-bases, the surface features of the bark and the occurrence of aerial roots capable of water absorption above the ground surface.

The proportion of water in the stem flow category is minimised by the more or less continuous 'umbrella' of the leaves and the form of the leaf blades. Numerous depressions and small water tanks can be formed by leaf-bases, particularly in vascular epiphytes and in trees developing terminal clusters of leaves with winged petioles, such as *Pycnocoma* spp. in African forests. By using radioactive tracers Lamont (1981) has shown that the aerial roots developed in the persistent leaf-bases of the grass-tree (*Kingia australis*) actually take up and translocate intercepted water. Thus precipitation which has never reached the ground surface can contribute to maintaining the internal water balance of this plant.

On the other hand, the amount of through-fall water (see Table 3.3) can reach about three-quarters of the precipitation, and seems to be enhanced by the abundance of drip-tips on the lower leaves of the trees, epiphytes and climbers (see section 4.2). According to Williamson *et al.* (1983), drip-tips homogeneously disperse much of the through-fall over the surface of the ground, and (by greatly reducing average droplet sizes) lessen

Table 3.3 Partitioning of precipitation (%) in some tropical forests. (After Doley 1981.)

Forest type	Location	Inter-ception	Stem flow	Through-fall	Reference
Rain forest	El Verde, Puerto Rico	27–38	0–1	62–73	Sollins and Drewry (1970)
Lowland dipterocarp forest	Jelebu, Malaysia	22	0.5	77.5	Manokaran (1979)
Natural teak forest	Lampang, Thailand	63	0	37	Chunkao *et al.* (1971)

the rate of soil erosion under a tree (see section 7.2). Between about a quarter and a half of the rainfall is directly evaporated, and it is this fraction that contributes most to the local and regional recycling of moisture mentioned above.

Another important source of moisture is dew, and during cloudless nights this can amount to 0.1–0.3 mm m^{-2}, equivalent to 100–300 ml m^{-2} of precipitation (Walter 1962). On the ground, dew is usually encountered only in larger clearings, for in the closed stand the level of radiant cooling and condensation is shifted upwards to the top of the canopy. The particular shape of the crown and the form, size and position of leaves may modify the rate of condensation. In tropical Africa, for instance, abundant dew is often formed on the large, digitate leaves of the 'umbrella tree' (*Musanga cecropioides*), and this can be seen dripping down to the ground. A similar occurrence can be seen in various *Cecropia* spp. of South and Central America. In other cases the droplets of dew may not be so large, and the leaves or leaflets may be smaller, so that the water does not reach the ground. However, its presence on the leaf surface for the first 2–3 h of the day can reduce transpiration and thus influence the daily water balance of the plant. It may also provide a potential water supply for invertebrates.

Fog and low clouds affect the ecology of tropical forests mainly in the mountains, and along rivers and the sea shore. Ellenberg (1959*a*) describes tropical fog-oases in the desert coastlands of Peru, where a low forest occurs at certain altitudes and the tree branches 'comb out' minute water droplets from clouds drifting in from the Pacific. Similar striking effects of cloud and fog are found in the mountains of East Africa and South-East Asia.

In the life of tropical forests, various aspects of atmospheric moisture have been correlated with plant growth and survival. The absolute amount of invisible water vapour has no particular ecological significance, but an important consideration is the relative humidity, the percentage of the maximum quantity of vapour which can be held by the air under a given temperature and pressure. Even more useful is the saturation vapour pressure deficit, measured in millibars or kPa, which takes account of the physical changes that make air of a given relative humidity much more drying if it is at 30 °C than at 20 °C. It can therefore provide a better estimate of how the atmospheric environment may affect the forest organisms. Table 3.4 shows how much various measures of humidity can change during a 24 hour period.

Numerous measurements have been made to find out how much truth there is in the idea of 'permanent humidity' in the tropical forest. The available data suggest that there is in fact great variation in space and time of this environmental factor. Day-time figures for relative humidity are particularly variable: high in the canopy during the mid-day period the level may fall to 70 per cent, while close to the ground in the forest interior it is generally above 90 per cent. In a submontane forest at El Verde, Puerto Rico, the records of hourly mean relative humidity at the top of the meteorological tower showed a decline during the afternoon to 80 per cent between January and July 1964, while at the ground in the same period the day-time depression of the mean hardly went below 95 per cent. For the October 1964 to April 1965 period the respective minimum limits were 90 and 95 per cent (Odum and Pigeon 1970).

In regions affected by seasonal rainfall fluctuations, day-time figures may be encountered which suggest very considerable though temporary dryness. The frequency and duration of such times of heightened water stress may determine the survival of species and thus the floristic composition of the forest; and also the rate of shoot growth through closing of stomata (section 5.2). During the 'Harmattan' season in African forests, the relative humidity of the air near the forest floor may fall almost to 70 per cent, a value so low as to damage the 'stenohydric' plants in the undergrowth (see Fig. 3.10). At a similar time of year, Cachan and Duval (1963) even recorded 30 per cent relative humidity of the air at 46 m near the tree tops. Similarly, Fetcher *et al.* (1985) have shown saturation vapour pressure deficits as high as 12 mb in the middle of the day in a clearing, with less extreme con-

Table 3.4 Diurnal fluctuation of humidity in tropical rain forest at Ankasa Forest Reserve, S W Ghana, during a spell of dry 'Harmattan' weather. Readings taken near the ground on 6–7 January 1967 in: (a) the forest interior; and (b) a forest clearing.

Time of day (h)	Relative humidity (%)		Saturation vapour pressure deficit (mb)		Dew point (° C)		Evaporative power of air* (ml dm^{-2} h^{-1})	
	(a)	(b)	(a)	(b)	(a)	(b)	(a)	(b)
20	94	87	1.6	3.4	**20.9**	20.5	0.7	0.7
22	95	91	1.3	2.4	20.8	20.4	0.7	0
24	91	94	2.2	1.4	19.6	19.5	0	0
02	96	94	0.9	1.4	19.6	19.3	0	0
04	93	93	1.7	1.7	**19.3**	19.1	0	0
06	**97**	**95**	**0.7**	**1.3**	**19.3**	**18.9**	0	0
08	92	94	1.9	1.4	19.7	19.7	0	2.1
10	85	70	4.4	10.4	20.7	20.4	0.7	8.5
12	71	51	9.3	22.1	19.4	19.6	1.4	**15.6**
14	**69**	**45**	**10.9**	**24.9**	**20.9**	17.9	**2.1**	14.9
16	71	60	10.4	16.0	20.8	**20.7**	**2.1**	8.5
18	77	75	7.4	8.0	20.8	20.2	1.4	2.8
Total (24 h)							9.1	53.1

(*measured by Piche evaporimeter with 3 cm diameter green disk.)

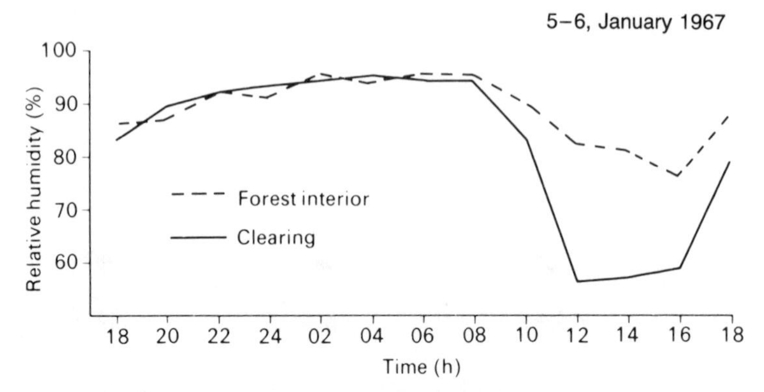

Fig. 3.10 Daily march of **relative humidity** in a rain forest in S W Ghana on an unusually dry day during a 'Harmattan' period. Both readings at 10 cm above the ground.

ditions in a gap. Conditions in the canopy were intermediate between the two, whereas there was only a slight increase in dryness in the undergrowth (Fig. 3.11).

All these figures show that plants and animals alike enjoy or perhaps endure in the upper canopy an entirely different daytime micro-climate from that in the undergrowth. Related to this may be the different size and perhaps shape of leaves of the same tree in various layers above the ground (see section 5.8), as well as the stratification of mammals at different layers (see Fig. 4.30).

In most cases there is a long nocturnal period during which relative humidity remains above 95 per cent and vapour pressure frequently approach the point of full saturation. The pertinent observations show a similar pattern during the night in all layers of the forest, including the ground layer of clearings, and it seems that the same applies even in drier seasons.

Measurements in the undergrowth of rain forests, using various kinds of evaporimeters, generally show very low rates of evaporation, chiefly because of the near-saturation of the air, the low radiation flux and the lack of wind near the ground. Naturally, there is usually little evaporation at night, and in the daytime near the ground it is usually at most about one quarter of the levels achieved above the canopy. Odum and Pigeon (1970) found negative values of transpiration near the forest floor at El Verde – a result explained in terms of presumed condensation of water vapour. In West Africa there are many diurnal periods during the rainy season without any marked evaporation near the ground, but during a dry spell rather high values may temporarily

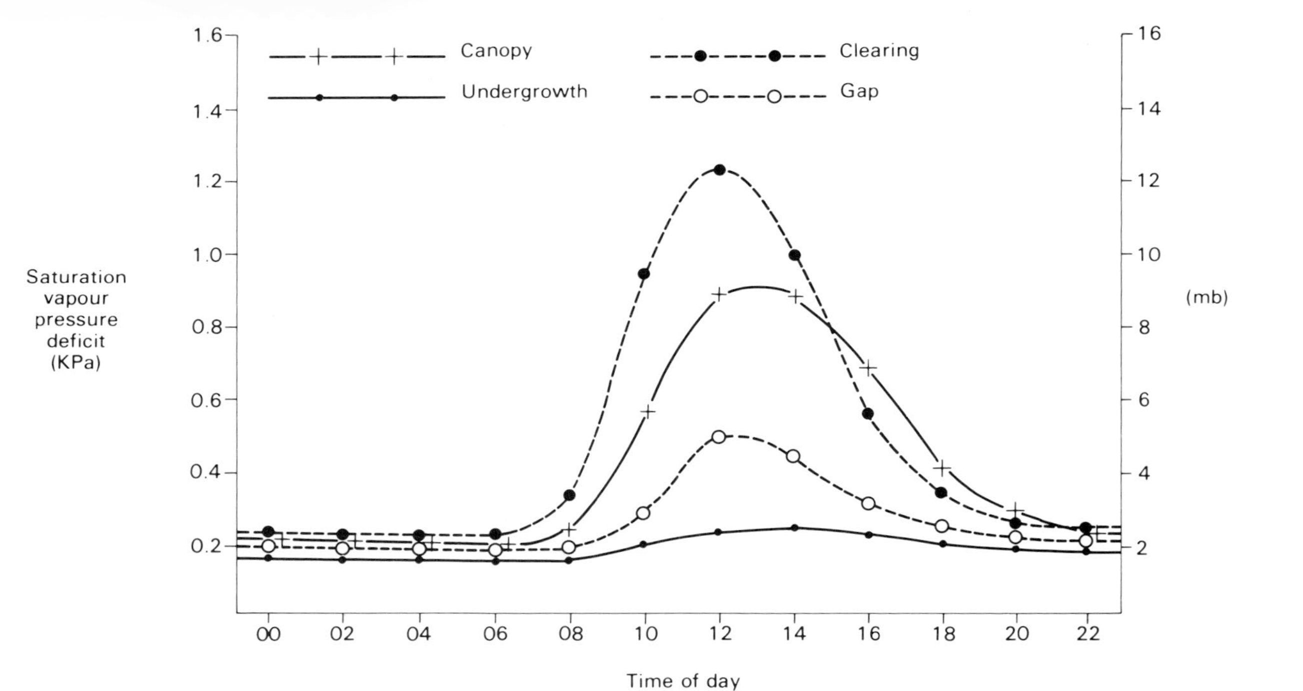

Fig. 3.11 Daily march of **saturation water vapour deficit** in the undergrowth, canopy, 0.04 ha gap and 0.5 ha clearing in a tropical lowland rain forest in Costa Rica. Geometric means of measurements made between April 28 and October 22 1981; higher readings showed greater variability than those around 0.2 kPa. (After Fetcher *et al*. 1985.)

be recorded (Table 3.4). In the upper tree layer, evaporation can even reach levels comparable to those occurring in the savanna (Jeník and Hall 1966). Both the crown leaves and epiphytes on the upper limbs generally possess morphological and physiological adaptations enabling them to survive the hot noon hours or exceptional dry spells (see Sinclair 1984).

Both the high water stress in the canopy and the low rates of evaporation in a continuously saturated atmosphere with little transpiration inside the forest cause physiological problems. For instance, to what extent does the resistance to water flow in the xylem of a tall tree accentuate the stress? And does the usual pattern of mineral nutrient transport in the xylem cease to be adequate without significant transpiration? Could the latter be a contributory factor to the slow and stunted growth of tree seedlings and pigmy trees?

If rainfall is the main factor controlling the geographical distribution of tropical forests, soil moisture is dominant in determining local patterns of forest types. Obviously position within the catena affects the degree of drainage, and consequently soil water contents in the rhizosphere. As shown in Fig. 2.6 (C), this includes sites with impeded drainage, free drainage and excessive drainage. Unless there are some irregularities in the soil profile – for example, the presence of an indurated hard-pan – habitats of impeded drainage are limited to the very bottoms of valleys and the flood-plains of large rivers. These waterlogged sites have distinctive floristic composition and overall forest structure: recognisable types are developed, such as tropical alluvial forest (riparian, occasionally flooded or seasonally flooded), tropical swamp forest, tropical evergreen peat forest, etc. (see classification in section 6.5). However, as already mentioned, the commonest forests are mesic forms, growing on slightly impeded and particularly on freely drained soils. Even during the height of the rainy season it is possible to walk in these forests without rubber boots, the conditions underfoot usually resembling those encountered, for example, in temperate beech forests.

Continuous measurements of soil moisture in tropical forests have been few. Schulz (1960) confirmed that soil moisture is strongly correlated with the texture of the soil. Bronchart (1963) recorded the variation in moisture in the surface layers of evergreen seasonal forest soil at 0° 30′ S in Zaire during a whole year (Fig. 3.12), and disclosed the surprising fact that water contents may drop below the permanent wilting percentage even in this vegetation type. In his study Bronchart also provided experimental evidence that fluctuation in soil moisture may control the seasonal flowering of the ground flora (see section 5.8).

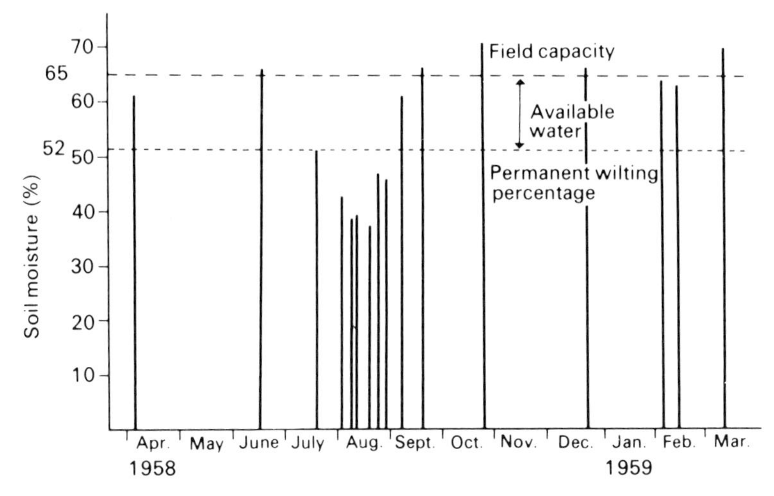

Fig. 3.12 Seasonal changes in soil moisture in an evergreen seasonal forest near the Equator in Zaire. Soil moisture expressed as % fresh weight: note levels well below permanent wilting % during August. Rainfall recorded at the same site varied from 0–125 mm per 10-day period. (After Bronchart 1963.)

Markedly different conditions for plant growth are found in waterlogged soils with impeded drainage. After periods of rain, these soils may be temporarily flooded, the duration of the flooding being from several hours to several weeks. Photographs showing tropical investigators walking knee-deep in water through the forest are almost certain to have been taken in the flooded bottom of a catena. The excess soil moisture interferes with aeration, another factor of paramount ecological significance, preventing many species of the mesic forest from growing in the alluvial complex of catenas.

Compared with mesic stands, alluvial, swamp and peat forests have fewer constituent tree species, with a tendency towards single-dominant stands. In Sarawak and Brunei, Whitmore (1975) describes a catena of various forest types from the edge to the centre of each peat swamp, where *Shorea albida* is a dominant tree reaching about 50 m height. Palms and bamboos are successful life-forms in swampy habitats in all parts of the tropics. Even in the herb layer large monocotyledonous plants predominate (Marantaceae, Zingiberaceae, grasses). Locally, the waterlogged site may be inimical to forest and patches of tall herbaceous vegetation can be found. However, some dicotyledonous

Fig. 3.13 Peg-roots of *Anthocleista nobilis*, a frequent type of pneumorrhiza in freshwater swampy soils. Easily overlooked because of their unobtrusive colouring and small size.

trees still flourish here, particularly if they have special adaptations for gaseous exchange (see Fig. 3.13).

At the other end of the moisture gradient, dry soils can develop locally, on the top of the catena or over excessively drained sands. Within these sites the forest is reduced in height, and drought-tolerant trees and grasses may occur in numbers. In some parts of the tropics one can even find savanna-like patches deep inside rain forest. In Zaire, these and the wetter patches called *esobe* are even found close to the Congo River (Germain, 1965), while in West Africa they can occur on indurated bauxite deposits capping very ancient hill-tops. However, in these and other cases man's interference could have played an important role in expanding the area or numbers of these savanna patches, because they have often served as centres for settlement and cultivation of crops.

Permanent or temporary excess of moisture inside tropical forests is a factor which can be correlated with various growth forms, environmentally induced modifications and adaptations. The shape of leaf blades is an interesting example of 'epharmonic convergence' in tropical forest trees, woody climbers and shrubs, whose leaves are very frequently extended into a narrow acumen called a 'drip-tip', which is usually curved downwards (Figs. 3.14 and 5.11). In a study in the Ankasa Forest Reserve, Ghana, we have found that more than 90 per cent of species in the forest undergrowth possessed leaves (or leaflets) with drip-tips. In an experiment comparing intact leaves with those deprived of the

Fig. 3.14 Prominent drip-tip on leaf of a sapling of *Dipterocarpus* sp. at Pasoh Forest Reserve, W Malaysia. The leaf has been attacked by leaf-mining insect larvae.

tip, the controls were absolutely dry only 20 min after rain, while de-tipped leaves were covered by spots of water even after 90 min. In conformity with the vertical environmental gradient, many emergent trees may develop drip-tips only in the undergrowth, while leaves occurring in the upper canopy are obtuse or truncate – for example, *Lophira alata* in Africa (see Fig. 5.20, and section 4.2).

Other hydrological peculiarities of tropical forests are the swamps and small 'water tanks' found in the upper canopy. They are formed amongst closely overlapping leaf-bases, both of trees and of herbaceous epiphytes. They may even contain specialised aquatic plants, like the carnivorous *Utricularia* spp. in South America, and commonly serve as breeding places for tree frogs and mosquitoes. Erect waxy flowers such as those of *Bombax buonopozense* may provide small temporary 'pools' which are visited by ants and other arboreal animals. *Homo sapiens* can even drink the potable water released when the slender tendril climber *Afrobrunnichia erecta* is cut (Hall and Swaine, 1981).

3.4 Availability of nutrients

The large biomass and great species diversity of mesic types of tropical forests may lead to two false assumptions: firstly, that the soil supporting such vegetation must necessarily contain a large supply of nutrients; and secondly that these reserves can be utilised for an unlimited period by agricultural crops, grazing land or plantations. However, the experience of local villagers is generally that yields quickly decline during two or three food-crop cycles within 1–2 years, suggesting a nutrient-poor status for most of the original forest soils. This has been confirmed by soil scientists and ecologists, for example, Nye and Greenland (1960), who gave the understandable explanation of the rapid weathering of soil minerals. The paradox of the luxuriance of the forest and the poverty of the soil has now been further explained by the discovery that numerous mechanisms have evolved that enable tropical trees and their associated forest organisms to capture and recycle the relatively restricted pool of elements in the soil solution (C. F. Jordan 1981), and those entering the ecosystem from the atmosphere.

Most organisms, as well as requiring certain major nutrients for the construction of their protoplasm, also to some extent reflect the chemistry of their current surroundings. The 'geochemical environment' of tropical forests varies a great deal, and more diversified still are the abilities of organisms in uptake, storage, transfer and release of elements. Taken together, however, this diversity seems to contribute to the remarkably efficient nutrient retention by intact ecosystems on many different soils. For instance, Stark and Jordan (1978) found that usually less than 0.1 per cent of the radioactive phosphate and calcium from solutions which they sprayed on to the floor of a Venezuelan tropical forest could be picked up in collectors placed below the rootmat and humus layer, even though water was definitely reaching them. Virtually all the tracers had been captured by the roots and soil organic matter (and perhaps also by mycorrhizal mycelia), and were being translocated out of the experimental plots without ever entering the mineral soil.

Another rather unexpected way in which plants might acquire nutrients has been clearly demonstrated for rice roots, and could perhaps occur in other genera. Rice plants thrived in water cultures which contained nitrogen only as organic macromolecules. Ultrastructural studies showed that these could apparently be taken up by the invagination of the plasmalemma to form vacuoles, rather in the fashion of the feeding of *Amoeba* (Nishizawa and Mori 1977). There is clearly much still to learn about the capacities of tropical root systems, and also fungal mycelia, but Golley *et al.* (1975) were already able to conclude from their analysis of several tropical forests in Panama that biological considerations override geochemical factors in governing the distribution and abundance of elements in the ecosystem.

Survival of a natural forest community is probably seldom primarily determined by nutrient availability (Golley 1983*a*: p. 139). However, there are some forests which achieve much greater productivity, biomass and species diversity than would be predicted from their nutrient-poor substrates. This subject is further considered in section 6.2, but it should be noted that complete clearance of such sites for farming deprives them of the adaptive nutrient conservation mechanisms which enabled the original vegetation to thrive.

Tropical soils are controlled by high levels of weathering, leaching, biological activity and input of elements by bulk precipitation. Favoured both by warmth and moisture, the strong weathering leads to deep soils that are rather uniform in their clay composition and overall chemistry. The clay fraction consists mainly of kaolinitic minerals with iron and aluminium oxides, and very few undecomposed minerals rich in nutrients are left within reach of plant roots. Moreover, the cation exchange capacity of these soils is very low, except for the thin layer of humus at the top. In high-rainfall regions oxisols are also highly acid, with soil reaction as low as pH 4. The C/N ratio tends to be high, with figures well over 10 being commonly found.

The nutrient cycle in the humid tropics cannot be properly assessed without an appreciation of the nutrients stored in the vegetation. Analysis of ferralsols has revealed the striking fact that the quantity of some exchangeable nutrients in the topsoil may only be of the same order as the amounts contained in the biomass of the living forest. Nye and Greenland (1960) undertook the laborious task of making a large number of analyses and calculations, a few of which are reproduced in Tables 3.5 and 6.3.

Table 3.5 Nutrient elements (in kg ha^{-1}) stored in the upper 30 cm layer of the soil, together with those stored in the biomass plus litter, in a mature secondary stand within a tropical rain forest. (After Nye and Greenland 1960.)

	N	P	K	Ca	Mg
Soil	1 830	125	820	2 520	345
Stand	4 580	12	650	2 580	370

It is important to realise that nutrient storage and also the accumulation of organic matter in tropical forest is very different from the equivalent processes in temperate woodland ecosystems (Fig. 3.15).

Once the tropical forest has been removed either by extensive extraction of timber or by burning and farming, the previous nutrient cycle is damaged and the humus content falls. Mineralised nutrients are quickly leached by rainfall, or washed away in surface run-off, and a considerable loss may also occur when agricultural crops are harvested. Altogether the soil conditions typically deteriorate, and farming generally ceases (see section 7.2).

The above account is, however, incomplete when long-term experience with shifting cultivation is taken into account. The bush fallow that develops during and after farming partly restores the fertility of the soil, or perhaps better of the ecosystem, though not to its previous level. Among the tree species involved in the succession some require a higher concentration of nutrients. The deep penetration of tree roots allowing the lifting of leached or newly weathered nutrients has been offered as an explanation for this point, and indeed, as pointed out later, tropical tree root systems are not as shallow as is often claimed.

Another factor in the restoration of nutrient resources is the input of elements by precipitation; though the annual amounts

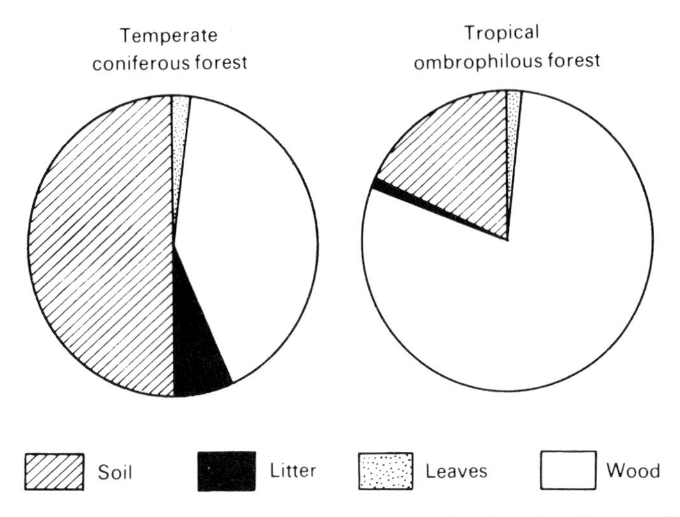

Fig. 3.15 Distribution of organic carbon in the **abiotic portion** (soil, litter) and **biomass** (leaves, wood) of tropical rain forest, compared with temperate coniferous forest. (After Kira and Shidei 1967.)

added may not appear to be substantial, in a decade the totals may nearly equal the annual flux due to litterfall (see Kellmann *et al.* 1982). Ten years, as it happens, is considered in many parts of the tropics to be the minimum period for bush fallow, before farming starts again. A further way in which more nutrients are captured and made available in forest soils is by the process of chelation. Many roots and micro-organisms appear to produce chelating agents, which bond reversibly with metal ions.

As in forests everywhere in the world, the cycling and *de novo* acquisition of nitrogen by the ecosystem play an important role, for it is a major element in all components of the forest. Table 3.6 shows an example of this dominance of nitrogen, and also the general distribution of dry weight and nutrients in various parts of the forest ecosystem. Among the many life-forms participating in the composition of tropical forests, there are a number of nitrogen-fixers that contribute substantially towards the nitrogen balance of the ecosystem.

Leguminous trees are fairly frequent throughout the tropics and consequently the importance of *Rhizobium* symbiosis may be anticipated. However, in excavations of roots in West African forests, many species of Papilionaceae, Caesalpiniaceae and Mimosaceae apparently had not developed root-nodules, although it should be noted that their production can sometimes be

Table 3.6 The elemental content in compartments of a nutrient-poor, lowland rain forest at San Carlos de Rio Negro, Venezuela. (From Herrera, sec. Golley 1983b)

Compartments	Dry weight t ha^{-1}	Nutrient content (kg ha^{-1})				
		N	P	K	Ca	Mg
Leaves	7	72	4	40	17	17
Stems	178	264	28	281	222	36
Roots	132	834	69	327	244	142
Fine litter	7	52	3	6	33	16
Dead timber	15	80	2	4	34	6
Surface soil						
0–21 cm	494	715	21	64	119	14
21–40 cm	1714	70	15	5	15	0
Subsoil						
40–75 cm	6072	0	18	4	61	4
Total (including soil 0–40 cm)	2546	2087	142	727	684	231

seasonal. In *Sesbania grandiflora*, *Gliricidia sepium* and *Leucaena leucocephala*, for instance, nodulation has been confirmed (NAS 1979; 1984), and is probably related to their capacity for rapid colonisation, even of degraded land (section 7.3). *Parasponia* spp. (Ulmaceae) also have functional *Rhizobium*-nodules (Akkermans *et al.* 1978).

Often overlooked are the other free-living and symbiotic micro-organisms which are found for instance in the rhizosphere and phyllosphere. Under aerobic conditions, the bacterium *Azotobacter* may be important, and *Beijerinckia* specifically in high-rainfall areas with very acid soils, while in anaerobic environments *Clostridium* spp. may take part in nitrogen fixation. Blue-green algae and fungi, epiphytic, endophytic and in the soil, are also involved; while actinomycetes such as *Frankia* form nitrogen-fixing associations for instance with *Casuarina*, an important genus for planting in adverse sites which is said to produce the best firewood in the world (NAS 1983*b*). Some of the small epiphyllous green algae, liverworts and mosses that grow on the surface of living leaves also fix atmospheric nitrogen.

The photosynthesis of the plants reaching the euphotic layer usually ensures a continuous supply of the energy-rich materials necessary for nitrogen fixation by the heterotrophs. Conversely, the nitrogen-fixing organisms help to maintain a balanced nutrient cycle. Fertiliser experiments tend to suggest that within a mature tropical forest no marked deficiency of nitrogen can be found, though in the same site under farming this is seldom true.

The soils in evergreen seasonal forest change their characteristics during the year. For instance, nitrates, the major source of nitrogen for plants, are very soluble and can be leached down the soil profile during the rainy season, where they may not be readily available; at the same time denitrification by bacterial activity may be further reducing the supply. In the dry season, however, especially towards its end, increased quantities of nitrates are often recorded in the topsoil. The role of lightning in fixing atmospheric nitrogen is not fully understood but it may make some contribution here. The improvement of nitrogen supply with the early rains may influence the growth of leaves, stems and fruits, which, as discussed in Chapter 5, is frequently taking place at this time. It clearly also has important implications for the time of sowing or planting of crops and trees.

Apparently related to the concentration of available nutrients in the surface layer, the roots of tropical trees often spread mainly in this part of the soil, though it is possible to exaggerate their shallow-rootedness. Excavations and semi-quantitative estimates in Ghana (Mensah and Jeník 1968) showed that the shape of the root skeleton of *Chlorophora excelsa* does not differ markedly from that of European oaks which serve as text-book examples of deep root systems. Tap-roots and large vertical sinkers can penetrate to a depth of at least 2 or 3 m. It is the distribution of small feeding roots which seems to be limited more strictly to the upper 5 cm layer of the topsoil (see Fig. 3.16 and Lawson *et al.* 1970). In temperate forests, most of the small feeding roots are to be found in a rather thicker layer of about

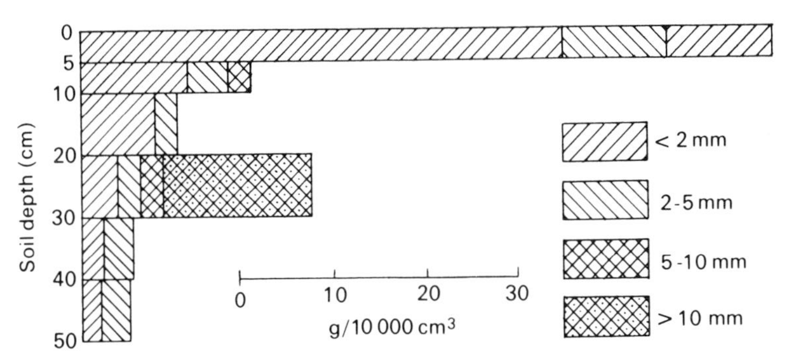

Fig. 3.16 Distribution of dry weight of tree roots at different depths in an evergreen seasonal forest in Ghana. Horizontal scale = weight of root material per unit volume of soil in 4 diameter classes.

10–20 cm depth, but in each case there is an obvious correlation with the humus and main nutrient source.

However, tropical soils may vary substantially from the common ultisols and oxisols. For example the andosols developed in regions occasionally covered by volcanic ash show a deep humus-rich and nutrient-rich topsoil, which offers a suitable environment for richly branched root systems. In the Andes, large areas of tropical montane and subalpine forest exploit the andosols, producing highly productive and species-rich stands.

As shown in Fig. 3.1, it is not unusual to find considerable quantities of crown humus in tropical forest. This consists of debris of all kinds: decaying leaves, twigs, bark, wood, fruits and animal remains, mixed with mineral soil brought up by ants from

Fig. 3.17 Crown of a large deciduous emergent tree in West Africa, heavily laden with epiphytes, including *Oleandra*, *Drynaria* and *Peperomia* (May 1967).

the ground. Crown humus accumulates particularly over large limbs spreading horizontally or obliquely, in forks in the trunk, on rough bark covering the boles, in terminal clusters of leaves and in pockets among leaf-bases. Epiphytes germinating in bark fissures or an initial pocket of soil may later enhance the accretion of humus and other particles (see Fig. 6.4). This can happen simply because the epiphytes provide a rough surface or because of special adaptations which increase the amount of humus trapped. Nest-fronds of some ferns (*Platycerium* spp., *Drynaria* spp.) and brush roots of some orchids (*Ansellia africana*, etc) are good examples of efficient humus collectors.

Gradually the upper surface of the bigger limbs (see Fig. 3.17) is covered by a thick carpet of crown humus, perhaps including the nests of invertebrates such as ants. This may be penetrated by the roots of the epiphytes, and sometimes by the roots of the host tree. That nutrients can in fact be absorbed by a tree's roots penetrating aerial pockets of soil on its own trunk has been demonstrated with radioactive tracers for the grass tree *Kingia australis* (Lamont 1981), and this has been termed 'translocatory shortcutting'. These features emphasise clearly the complex structure of tropical forests, which provide nutrients not only below ground level, but at any height above the ground.

3.5 Disturbances in the forest

It is not generally realised that, in addition to the profound effects that past climatic changes have had upon tropical forests (sections 2.2 and 4.5), they are subject today to a range of powerful disturbing forces. The trees, vertebrates and other forest organisms are adapted to the conditions that generally prevail in the locality, but are also subject to rarer events that bring sudden stress. For example, as indicated in section 3.2, an occasional drop in temperature to levels where chilling injury starts to occur could be as significant ecologically as the normally prevailing conditions to plants growing at the foot of large mountain ranges.

Indirectly, **wind** influences the geographical distribution of tropical forests through the associated movement of moist and dry air masses, which broadly determine climate and seasonal changes. Departures from the usual patterns may however occur at irregular intervals and bring unusually dry conditions even into the most equable and perhumid of equatorial climates. On the sea-shore, salt-laden winds may be responsible for the presence of low littoral scrub, gradually merging into normal rain forest.

Similarly, the dwarfing of the forest and the presence of flag-shaped trees on the ridges and tops of wind-swept mountains can be attributed to the mechanical and physiological effects of wind, the latter including a pronounced increase in the drying power of the air. Some exposed hilly sites may be covered not by forest, but by an open type of vegetation, without large trees.

From the point of view of the internal organisation of tropical forests, even occasional breakage of crowns and sporadic uprooting of trees can have far-reaching implications in the life of the forest community (see section 6.4). Wind interferes profoundly with the forest equilibrium, changing at a stroke the conditions for regeneration, growth and reproduction in the lower layers. The clearing caused by a single wind-thrown tree may spread over half a hectare, since other nearby trees linked by climbers can be dragged down or snapped. Due to this factor, many heliophilous species are able to survive in large blocks of closed forest, often as dormant seeds in the soil. Thus wind plays an important role in colonisation, succession and speciation in tropical forests by causing gaps and creating specialised habitats. It also contributes to these processes through assisting in pollination, and in seed and spore dispersal. However, bearing in mind that calm air usually prevails in the interior of the forest, it is not surprising that other agencies play a greater role, with wind being important especially in species which reach the upper tree layer.

The most pronounced effects of wind are of course those of severe storms. In some regions, these appear to be relatively frequent, and the tracks where they have passed may be clearly distinguishable by their different structure and floristic composition. Thus for example the effects of tornadoes have been studied in Nigeria, hurricane forest described in the West Indies, storm forest in Malaysia and cyclone scrub in north-east Australia. Wind-speeds during these violent storms may occasionally reach 150 km h^{-1}, and so large limbs on emergent trees can be broken, many trees uprooted and even the whole of the upper canopy defoliated. Naturally, therefore, the effect of repeated storms overrides the usual ecological factors of soil and microclimate, so that instead of a stable forest climax a disturbed community is found, with trees and lianes lying helter-skelter, and vines, herbs and pioneer trees becoming dominant.

Fire is generally associated in people's minds with man, and indeed it has been and is a very potent tool with which the forest is changed: to cultivate food crops, clear land for grazing animals, or protect the village against encroachment by the forest. However, fires can also occur through natural causes, notably from lightning strikes and along the edges of lava flows and from hot ash from active volcanoes. In some parts of South-East Asia and East Africa, it is possible that the disturbing effects of such natural fires, together with the nutrient-rich soils derived from old lava flows, may have contributed significantly to the richness of the tropical flora.

Natural fires, or those caused accidentally or intentionally by man, cannot normally spread far into undamaged mature forest, because in most cases there is insufficient dry, combustible material on the ground to spread the flames, still less to sustain a crown fire. However, severe and prolonged droughts can apparently change the situation, as in some Indonesian and Malaysian parts of north-eastern Borneo where rainfall from July to November 1982 was more than 60 per cent below normal, while almost no rain at all fell between February and mid-May 1983. Many evergreen species shed leaves, and with the low atmospheric humidities and drying winds there was an accumulation of dry litter on the forest floor (Malingreau *et al.* 1985).

Fires, generally triggered by the burning of adjoining land that was being cleared for agriculture, ran through about 800 000 ha of primary lowland rain forest in East Kalimantan, killing some but not all of the trees. Many of the food sources for animals were lost, as well as the shelter that allowed them to escape from hunters and animal predators. The fires were so extensive that they could be mapped by remote sensing from satellites. The destructive effects on plants and animals was most severe in peat-swamp forest that had dried out enough to allow the organic matter to be set on fire. Swamp forest in Brazil has also been known to be set alight occasionally in the same manner.

When the forest has been opened up in various ways, so that there is not a full stocking of trees, or there is young secondary re-growth, there is usually more material near the ground to burn. Over 2 million ha of such areas were burnt in the same fires in Indonesia, while after a severe drought in Ghana in 1983 it was estimated that as much as 50 per cent of the growing stock in the forest reserves had been damaged by fire. The likelihood of an area catching fire generally increases toward the drier margins of the seasonal forests, while teak forests and plantations have a particularly high fire risk because of the large, dry leaves on the ground. Beyond a certain point, the incidence of fires upon a particular site will lead to permanent loss of fire-tender tree species, and the invasion of fire-tolerant grasses such as *Imperata cylindrica*. Where there are other stresses as well, the forest vegetation may change to savanna or grassland, or to scrub or even bare rock (see Fig. 2.11).

Large volcanic eruptions can disturb not only the local vegetation, but forests throughout the world, because of the very large amounts of dust lofted high into the atmosphere. The screening of sunlight can be sufficient to lower regional temperatures by 5–8 °C (Robock and Mass 1982), and global averages by as much as 1–2 °C (Kelly and Sear 1984), the effects lasting for months. Such temperature drops are large enough not only to have direct effects on the forest plants and animals, but also to influence them indirectly by temporarily altering the movement of air masses which determine climate and weather. Lightning strikes are quite frequent during tropical storms in some localities, while earthquakes, tidal waves and other exceptional floods are examples of uncommon but powerful disturbing factors that can occasionally devastate the forest, producing entirely new surfaces for colonisation. An unusual opportunity to study recolonisation was provided by the explosion of Krakatoa in 1883 which produced islands that were initially devoid of plant and animal life (Thornton 1984).

A different kind of damage occurs when an increase in the population size of a species becomes sufficiently large to lead the outbreak of a pest or disease. Interestingly, the cases which have been documented for natural tropical forest all appear to concern gregarious host populations, not the more usual mixed stands (UNESCO 1978). For example, sandal spike disease, probably caused by a virus, has so damaged *Santalum album* in India that future supplies of the extremely valuable sandalwood oil are uncertain. Single species plantations tend to be more susceptible than secondary forest to insect pests such as psyllids or the grasshopper *Zonocerus* sp. Sometimes it may be impossible to establish pure stands, as for instance of *Chlorophora* and some Meliaceae in West Africa, because of crippling attacks by shoot-borers, although the same species are not infrequent in the primary vegetation. It seems that the diversity of most natural forests may confer a measure of resilience against an outbreak, since the chance of the pest or pathogen spreading rapidly is greatly reduced by the low population density of the potential host (see also *Atta* spp. in section 4.4).

Finally, the effects of two species of mammal should not be forgotten. Elephants browsing, breaking down branches and uprooting trees are a potent source of gaps, thus stimulating regeneration. Like most other herbivores they eat selectively, and thus may depress the numbers of particular species, sometimes those that are economically important. The widespread effects of *Homo sapiens* are discussed in section 2.3 and Chapter 7. This species alone has the capacity to disturb and even eliminate all the tropical forests in the world, and thus is a potential threat to all their other biota. On the other hand, it also possesses the unique ability to decide to base its individual and group activities both on the experience of the past and prediction of the future.

4 The forest community

Forests all over the world are typically complex communities of species, many of which may remain unknown to their investigators. Generally only the vascular plants, vertebrates and the commoner insects are readily recognised by ecologists, and a large team of narrowly specialised taxonomists would be needed to prepare a complete list of species in a particular locality. For a full inventory of numbers, approximate estimates generally have to be used, while simplified models are perhaps the best solution to the problems of looking carefully at quantitative relationships. All these considerations apply with added force to tropical forest communities, particularly for the species-rich stands developed in moderate, mesic habitats, where evolutionary processes have provided the lower latitudes with a much richer biota (see section 4.5). Moreover, due to the historical development of science, knowledge about the members of tropical forest communities still lags behind that concerning temperate woodlands.

At their first approach to tropical forest ecology, students may be misled by the apparently confused tangle of trees, climbers and epiphytes occurring on river banks, at roadsides, or in the neighbourhood of villages. The 'inaccessible jungle' is, for the most part, a human-induced marginal community (Fig. 1.2). Another incorrect generalisation may arise because of the usual emphasis placed upon the prevailing species-rich, mixed stands that are developed as a climax over mesic tropical soils, or on the fertile flood-plains of the less turbulent rivers. However, the stress factors found in the intertidal zone, or over excessively dry or particularly nutrient-poor soils tend to produce very much simpler forest communities, often dominated by a single tree species, with fairly monotonous associated organisms. For example, patches of mangrove woodlands, tropical swamp and peat forests or tropical heath woodlands can occasionally be surprisingly uniform in their species composition.

In order to grasp the essence of a forest community such as the one shown in Fig. 4.1, one has to appreciate the role of the various populations of woody plants, for their contributions towards its biomass and interspecific competition are clearly

Fig. 4.1 Lowland dipterocarp forest near Kuala Lumpur, Malaysia (November 1980).

essential. Unlike the situation in temperate forests, unified regulations for mensuration of timber trees have seldom been developed in the tropics. Instead, closer relationships between ecology and forestry have led to the development of special sampling and enumeration techniques (Taylor 1962). Strip survey, line-plot methods and nowadays aerial photography are used to assess the growing-stock of timber. Detailed mapping of crown projections and of vertical profile diagrams (Figure 4.2) is done according to a variety of rules that define the size of the plot and the minimum girth of woody plants to be included in the census (see Richards *et al.* 1939; 1940). Later on, numerous parameters of primary production and sampling procedures of trees have been used (Ogawa *et al.* 1965; UNESCO 1978).

Fig. 4.2 Vertical diagram of a tropical rain forest in the Ankasa Forest Reserve, S W Ghana. Width of the sample strip was 7.6 m. (Jeník and Hall, unpublished material.)

Moreover, the other life-forms of plants in the forest also require adequate enumeration and sampling, which can be a difficult task in a species-rich and epiphyte-rich tropical community. Unlike the case with the well-described woodlands of Europe or North America, it is seldom possible to execute a complete phytosociological *relevé* over a satisfactory sample plot area in a tropical forest community. A single *relevé* on one 25 × 25 m sized plot may need a whole day of field work plus additional research in the laboratory or herbarium. Hall and Swaine (1981) needed four years to assemble lists of the vascular plants in 155 sample plots measuring 25 × 25 m for their comprehensive survey of forest communities in Ghana, West Africa. A simple descriptive field technique in the style of the Zurich–Montpellier School (Becking 1957) cannot be adopted because of difficulties in the analytical phase, namely:

1. The frequent lack of apparent spatial pattern, preventing the delimitation of homogeneous sample plots and the definition of meaningful layering;
2. Trouble with the on-site identification of plant species, and assignment of semiquantitative 'cover' values to particular plant populations. Various 'quantitative' parameters of plant communities (see review by Shimwell 1971) can be applied only in detailed studies limited to a single location.

Besides the vascular plants that create the bulk of its primary production, other life-forms are indispensable in the forest community. Bacteria, cyanobacteria, fungi, invertebrates and vertebrates require special field techniques in order to overcome intrinsic difficulties in sampling them. Problems arise for example from their microscopic size and high numbers; inconspicuousness because of their rarity, hidden or nocturnal life; ephemeral occurrence; inaccessibility in the above-ground or underground space; etc. However, although sometimes very abundant, these microscopic creatures are not necessarily rich in species.

4.1 Three-dimensional structure

The forest community is a three-dimensional phenomenon. The location of any member of this community, be it an individual, its various organs or an entire population, can therefore be fully described only by a system of Cartesian co-ordinates in space. Three orthogonal projections would normally be needed in order to illustrate its position and structure, height, width and depth. However, two projection planes are often used in the analysis of a forest community, based on the negative or positive geotropic orientation of main stems and roots relative to a flat ground surface. Accordingly one may speak in general terms of the vertical or the horizontal structure of the forest.

Amongst the features of **vertical structure**, four characteristic points deserve special attention:

1. The maximum height achieved by the canopy and emergent trees.
2. The above-ground layering of shoots and dependent life-forms.

3. The underground layering of roots, rhizomes and dependent edaphon.
4. The maximum depth reached by the roots.

Although very large, the biggest emergents of the tropical forest are not quite as tall as, for example, the redwoods of California or some of the *Eucalyptus* trees of subtropical Australia. The maximum height of the canopy in tropical forests seldom exceeds 50 m, though exceptional specimens may reach 70 or 80 m. As shown in Figs 2.4, 2.6 and 2.7, the general height of tropical forests is influenced by the amount and periodicity of rainfall, and by temperature, soil drainage and nutrient levels. On the whole, the tallest structures are found in areas with an alternation of humid and drier seasons, rainfall totals of about 2 000 mm and freely drained soils. These conditions presumably favour in various ways the growth and longevity of the trees. By contrast, very high rainfall areas without marked seasonal changes and sites with impeded drainage often tend to produce rather smaller trees, sometimes with a shorter life-span. Other adverse environments which may reduce the forest height are found, for example, on wind-swept slopes and hilltops, seashores influenced by salt-laden winds and in mountainous areas (see section 3.5).

The most luxuriant trees and stands of all, sometimes exceeding 60 m, are to be found in the upper freely-drained parts of the flood-plains of rivers, where flooding leaves fertile silt, and temporary water stress can occur. *Ceiba pentandra* is one of the giant trees that protrude above the alluvial forest in many tropical regions.

In the Amazon Basin, Hueck (1966) found the average height of the forest to be 30–40 m, with *Dinizia excelsa* (Leguminosae) and the Brazil nut, *Bertholletia excelsa* (Lecythidaceae) as examples of species where individuals may reach 50 m. In tropical Africa similar size ranges are encountered, and a gradient of increasing forest height can be recorded along an ecotone from high-rainfall areas (stands approximately 30 m high) towards evergreen seasonal forest (about 40 m). *Entandrophragma cylindricum* (Meliaceae) and *Piptadeniastrum africanum* (Leguminosae) are examples of giant trees which occasionally protrude far above the average level of the canopy. In the Indo-Malaysian tropical forest region, *Dryobalanops aromatica* and other species of Dipterocarpaceae can reach 60 m, and the tallest tree ever recorded in the tropical forest grew here: a specimen of *Koompassia excelsa* (Leguminosae) which measured 84 m (Foxworthy 1927). Altitude also influences the vertical structure of tropical forests, as can be clearly seen from Table 4.1.

Table 4.1 Relationship between altitude and the height of the forest canopy in W Malaysia. (After Robbins and Wyatt-Smith 1964)

Forest formation	Altitude (m)	Height (m)
Lowland dipterocarp forest	150	42
Upper dipterocarp forest	800	30
Lower montane oak–laurel forest	1 500	21
Montane ericaceous forest	1 800	15

The top surface of the canopy in tropical forests is undulating, chiefly because of the emergent trees of the upper tree layer (see Fig. 4.1). When viewed from an aeroplane the distinctive appearance of untouched tropical forest has been described as a 'cabbage patch', because of a superficial resemblance to a field of closely planted *Brassica*. Most trees, of course, never become emergent even at maturity. The taller ones create the continuous canopy of the middle tree layer, a prominent boundary between the euphotic and oligophotic zones of the forest. Lower down there are two classes of smaller trees: some seedlings and saplings which may later reach the higher levels, and other 'pigmy trees' which stay permanently in the undergrowth.

Interpretation of the term 'layering' has caused considerable controversy among ecologists. Some hold that marked vertical stratification is very characteristic of tropical forests, while others state that no clear-cut layering can be found. Richards (1983) has recently pointed out that single-dominant tropical forests show clearly defined strata, but mixed forests usually do not (Fig. 4.3). The reasons for disagreement lie in how strictly the terms layer or stratum are defined. In their pure sense they include the idea of discrete horizons filled with different sets of branches, foliage or roots. In this respect, there are no genuine layers in mixed tropical forests, and only weakly manifested layers in species-poor stands. More commonly, however, the term layering is used to divide the vertical structure into convenient sectors (Table 4.2), occupied by more conspicuous or more aggregated organs of plants, and often occupied by different animals (see Fig. 4.31).

The complexity of the above-ground vertical structure in tropical forests is greatly enhanced by the presence of numerous epiphytes and climbers. In the montane and subalpine belt, macro-epiphytes and micro-epiphytes (see section 4.3) are particularly abundant, and small aerial patches of foliage and roots ('gardens') situated at various heights above the ground deny the common organisation of phyllosphere and rhizosphere observed

Fig. 4.3 Freshly cut vertical profile through an evergreen seasonal forest at Kade, Ghana, showing layering, emergents and climbers of a primary stand.

Table 4.2 Theoretical layering in tropical forest, based on West African rain forest and evergreen seasonal forest

Layer	Brief description
Upper tree layer	Emergent trees, woody climbers and epiphytes above 25 m
Middle tree layer	Large trees and woody climbers from about 10 to 25 m
Lower tree layer	Small trees and saplings reaching about 5 to 10 m
Shrub layer	Tree seedlings, shrubs, *krummholz*, small pigmy trees from 1 to 5 m
Herb layer	Smaller tree seedlings, forbs, graminoid plants, ferns and bryophytes up to 1 m
Upper root layer	Compact root-mass in the surface soil down to 5 cm
Middle root layer	Less abundant tree roots in the subsoil from about 5 to 50 cm
Lower root layer	Scattered tree roots below 50 cm

in terrestrial vegetation. In forests of the montane belt of the Andes, for example, populations of Orchidaceae, Bromeliaceae and Araceae locally show that the ground level can be an obviously less important substrate for herbaceous plants than the trunks, branches and leaves at various heights above the ground. Epiphytes germinating first in the tree canopy and later spreading their roots toward the ground are also an unusual phenomenon which disturbs the idea of stabilised layering (see Frontispiece). At different times in its life, a liane can develop its stem, twigs and foliage anywhere from the herb layer up to the top of the emergents, as it climbs upward on trees. It may later slip or fall down if its support should break or rot, forming great loops on the ground and sending its shoots up again by another route.

Genuine shrubs (woody plants, branching at the base) are seldom found in the untouched forest. The majority of the small woody plants scattered in the undergrowth are either seedlings of bigger trees, palms or trees *en miniature* called pigmy trees. The last mentioned have a distinct, often unbranched axis, with small cluster of leaves, as shown by many species of *Dracaena* (see Fig. 6.4) and *Pycnocoma* (see cover). In the West African forests *Sloetiopsis usambarensis* and *Scaphopetalum amoenum* develop vegetatively spread thickets classified as a special life-form called *krummholz* (Jeník 1969); arched stems produce adventitious roots upon touching the ground, and successively develop at the top of the loop lateral branches that repeatedly arch and root into the wet soil. How this may be controlled is one example of the fascinating physiological problems of apical dominance that recur with tropical woody species.

The herb layer in closed-canopy forests seldom covers more than 10 per cent of the surface, and is thus comparable in extent with the typical undergrowth in the shade of European beech forests. There are also generally fewer species of ground herbs than trees and crown epiphytes. For example, in an area of a few square kilometres of primary forest in Guyana, the herb layer comprised about 30 species of flowering plants and 10–20 species of ferns, as compared with several hundreds of species of large woody plants (Richards 1952).

The interface between above-ground and underground space in tropical forests can be rather obscured by the loose layer of litter and freely growing roots, sometimes forming a mat of superficial organs exposed to the moist atmosphere. Data on the underground layering of roots and the depth of root penetration are still scarce. As discussed in section 3.4 and shown in Fig. 3.16 the available evidence suggests that the upper root layer may be very distinct, and that, contrary to the casual impression of general shallow-rootedness, both the middle and also the lower root layer may be represented. Some tropical trees can form large tap-roots and sinkers penetrating down to a depth of several metres, though the root systems of climbers, pigmy trees and ground herbs are usually confined to the upper root layer. In the waterlogged soils of alluvial, swamp and peat forests, the prevailing depth of rooting is generally shallow, deeper root systems apparently occurring only if supported by special root adaptations, such as peg-roots and knee-roots (see section 4.2).

The **horizontal structure** is a complicated phenomenon and its analysis should cover various layers above the ground and in the soil. Obviously, the position of the tree boles and tree canopy determine the general patterns of shoots, roots and animal life. Consequently, the first distinction concerning horizontal structure to be made is that between the closed or closed-canopy forest (*forêt dense*) and open forest (*forêt claire*). Another important consideration is the horizontal distribution and size of clearings or gaps devoid of trees, and the occurrence of patches of thickets or dense groups of larger trees resulting from natural regrowth in previous gaps, or from colonisation of laterally illuminated forest margins.

Applying the scheme used in mixed temperate forests, Whitmore (1975) described the forest growth cycle and its reflection in the forest mosaic as forming three phases: the gap phase, the building phase, and the mature phase. In another approach, the mature phase can be further divided and a *disintegration phase* set apart. These phases do not remain static but change successively with time, and their delimitation may vary slightly, according to the growth cycles and ecological 'strategies' of the dominant tree species. In general terms, the phases are distinguished as follows:

1. The **mature phase** (corresponding to the dim phase of illumination near to the ground – see section 3.1) is the most extensive and relatively most stable part of the forest mosaic, provided that extreme physical disturbances and large population explosions are absent. Under the unbroken canopy only a low density of sciophilous herbs, seedlings and pigmy trees survives. It is usually easy to walk around (see Fig. 1.3). If interspersed by more numerous old, moribund trees this is termed the *overmature phase* (Brünig 1976), and is marked by breakage of individual limbs and boles, by enlargement of the remaining healthy trees, and by luxuriant growth of epiphytes over large limbs.
2. The **gap phase** (corresponding to the light phase with regard to illumination) may be created by the falling down of

individual large trees, or when a group of trees is broken down by the falling crown or by interconnecting climbers. Natural decay by fungi, destruction by insects or mammals, and windthrow by hurricanes are amongst other factors causing such openings. The interruption of the canopy stimulates luxuriant development of climbers and ground herbs, and encourages faster growth of existing tree seedlings and saplings (Fig. 4.4). Coppice shoots sprout rapidly and dormant seeds, particularly of pioneer species, are stimulated to germinate (see sections 3.1 and 5.11). Unless there is abnormal wind action, the area of these natural gaps usually represents less than one-twentieth of the entire forest. In swamp forests this phase is more extensive because the trees tend to die earlier: large forbs, grasses, ferns and palms spring up most frequently in the gaps.

3. The **building phase** (corresponding to the dark phase) develops as a successional stage in the undergrowth, as thickets of saplings and groups of young trees fill up the gap. Since the plants are often growing close together and rapidly, this phase may possess the highest production of biomass.

Fig. 4.4 Interior of a W African rain forest, showing relatively abundant suppressed seedlings, saplings and pigmy trees in the gap phase. Note also the rotting logs on the forest floor, sign of a previous *chablis*.

> Pioneer species often sprout readily from the seed stock and from seed rain, but the climax tree species may steadily follow. However, the building phase can sometimes be retarded or even inhibited by the expansion of vegetatively spreading bamboo thickets, or by tree-ferns, palms or competitive stands of *krummholz*. All these kinds of building phases create unfavourable conditions for herbs and seedlings at the ground layer.

Another approach to studying variation of the three-dimensional structure of tropical forest, proposed by Oldeman (1978), is described in section 6.4.

Forest margins represent another interesting feature in the horizontal structure of the tropical forest (Fig. 4.5). Natural

Fig. 4.5 Rapid regrowth of thicket on forest margin in W Africa, composed chiefly of *Musanga cecropioides*. (About 3 years since clearing of adjacent forest.)

margins are formed along riversides (Fig. 4.6), on the seashore and around large clearings caused by hurricanes. Man-induced edges to the forest occur, for example, along roadsides (Fig. 1.2), around compounds and villages, and on sharp fire-induced boundaries between forest and savanna. These marginal communities are quite unlike those in the interior of the forest (Fig. 1.3). With more small trees, vines, woody climbers and heliophilous ground herbs, a dense strip or border is created. The luxuriance of this marginal thicket is presumably due both to increased light intensity and decreased root competition, and may sometimes be enhanced by nutrient release, river fogs or spray. Birds, small mammals and insect colonies may also contribute, by their dispersal of seeds or burrowing activity, to the different constitution of the forest margins. How quickly and efficiently such a thicket appears in the humid tropics can easily be observed along newly constructed roads.

In American tropical forests *Cecropia* spp. create dense marginal regrowth, and similarly in Africa *Musanga cecropioides* quickly germinates and predominates in the roadside thicket. The

Fig. 4.6 The Río Quijos, a headwater of the R. Amazon, in full flood, creating fresh forest margins. Note pioneer *Tessaria* trees protruding above the water.

latter species can achieve 10 m in height and 30 cm in girth within 5 years (see also Fig. 4.5), and its reliable re-appearance is due to the extensive seed bank accumulated during the preceding years of the mature forest (Whitmore 1983; see also section 5.11).

How trees are distributed within an area with relatively uniform soil, topography and climate is probably the aspect of horizontal structure that has been most studied. To begin with, investigators may ignore species, and just take into account the diameter and/or height of the trees. For example, Cain *et al.* (1956) counted trees in the Amazonian forests having stems exceeding 0.1 m diameter at breast height (dbh), while Klinge (1973) recorded the numbers of all plants exceeding 10 cm in height, dividing them into six layers with different crown heights (Table 4.3). In a detailed analysis of forests in the Jengka Forest Reserve, Western Malaysia, Poore (1968) concentrated on trees that reached the canopy and were over 3 ft in girth (0.3 m dbh).

Consequently there is considerable variation in the way that values such as 'abundance' and 'density' have been estimated, and even calculations of diameter class and basal area (see Table 6.2) may not be strictly comparable between different studies. Nevertheless, it can be seen from Table 4.3 that the density (in basal area per hectare) of the 50 emergent trees was 7.1 m^2 ha^{-1} while that of the 1 075 middle tree layer trees totalled 19.5 m^2 ha^{-1} (Klinge 1973). For comparison, the total basal area of trees in natural forest in the Central European mountains varies from 40–50 m^2 ha^{-1} (Pruša 1985).

Table 4.3 Layering, density and basal area (all plants) in the Central Amazonian rain forest. (After Klinge 1973)

Layer	Average crown height (m)	Number of individuals per hectare			Basal area (m^2 ha^{-1})
		Trees	Palms	Total	
A	35.4–23.7	50	0	50	7.1
B	25.9–16.7	315	0	315	14.6
C_1	14.5– 8.4	760	15	775	5
C_2	5.9– 3.6	2 765	155	2 920	2
D	3 – 1.7	5 265	805	6 070	1
E	1 – 0.1	—	—	83 650	1
Totals				93 780	30.7

Very important for the understanding of the complete three-dimensional structure of the forest are studies that deal with the distribution of individual species, associations between species, patterns of dispersion and of girth classes, and various parameters or indices of diversity. These require a great deal of work. because of the large areas involved. Hubbell and Foster (1983) report the complete mapping and identification of all trees over 0.2 m dbh in a 50 ha plot on Barro Colorado Island, Panama, which revealed patterns of tree distribution and abundance. Most species were patchily distributed, some were randomly scattered, and few if any were uniformly spaced. Among the patchily distributed species were several that followed closely the topographic features of the study area. Most canopy species, however, had their own individualistic dispersion pattern, with only a weakly developed association with habitat in some instances. There are some signs that it may become possible in such investigations to relate species distribution to such factors as horizontal variation in a soil nutrient, or previous positions of large trees, termite mounds, etc.

4.2 Diversity of tree form

In analysing the members of the tropical forest community, one cannot realistically adopt any narrow and rigid definition of a tree, such as those used by foresters in the temperate zone. Certainly, 'woody plants showing a single erect trunk more than 7 cm dbh and at least 5 m in height' are a prevailing life-form in tropical forests. However, there are also a number of other tree forms that, by different morphological 'strategies', attain a comparable size and play a similar role in the ecosystem as do dicotyledonous and coniferous trees (Hallé *et al.* 1978). The large palm family (see section 4.3) make a considerable contribution towards these additional tree forms – for example, every tenth tree between 1.5 and 38.1 m in height was a palm in a central Amazonian rain forest (Klinge 1973). Tree-ferns, screw-pines and bamboos are amongst the other unconventional 'trees' which help to characterise tropical vegetation, and therefore merit special attention.

Many attributes of dicotyledonous tropical trees do not differ from those in other parts of the world, whereas there are certain features of branching habit, foliage, flowers, fruits and root systems which are seldom or never encountered in other geographical zones. It is to these that the following pages chiefly refer.

First a general account of root behaviour will be given, since this is less well known than that of the shoots. In describing the form and growth of roots, one has to bear in mind both ontogenetic changes and the interaction between heredity and environmental factors. Thus one may expect that the growth pattern of a large skeleton root near the base of the bole will diverge sharply from that of a terminal root at the periphery of a richly branched root system. Additionally, the form of roots that are constrained by a hard lateritic pan or a permanently impeded horizon needs to be distinguished from the types of root system that might have developed had the trees been growing in deeper and more fertile soil.

Aerial stilt-roots (Jeník 1973) are prominent examples of a special feature of tropical forest trees, and have therefore frequently been photographed. However, many other tropical trees develop some part of their root system above the ground. There is a great variety of these roots including small spur roots, large buttresses, and numerous kinds of pneumorrhizae. All these features provide valuable characters for identification of tree species in the field (see, e.g., Schnell 1950).

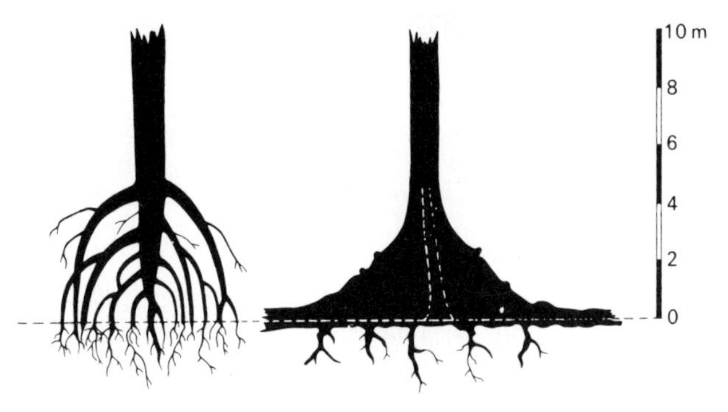

Fig. 4.7 Stilt-roots of *Uapaca* sp. (left); and heavy **buttresses** of *Piptadeniastrum africanum*.

Buttresses (Fig. 4.7) vary in size, shape and number, at their least prominent being merely a small spur or swelling at the base of the stem, not unlike those found in temperate conifers. At the opposite extreme they may develop as massive structures reaching as much as 10 m up the stem and much further in the horizontal plane. In such cases they provide a considerable obstacle to felling, so that a special platform may have to be built. Examples of emergent trees possessing very large buttresses are *Mora excelsa* in South America, *Kostermansia malayana* in South-East Asia and *Piptadeniastrum africanum* in Africa.

On some trees the buttresses are very thin and may sometimes be used directly as boards for house building. Others are very twisted, and anastomose providing unusual microhabitats. From the morphological point of view they are generally part root and part stem. The function of buttresses of different types is not clearly understood, although many contradictory theories have been offered. It seems likely, however, that some of them represent adaptations increasing the stability of big trees. Evidently their high surface/volume ratio, emerging through natural selection in the tropics, was largely unsuited to the seasons at higher latitudes (Smith 1972).

In the lower portion of the trunk of some species it is normal for adventitious roots to be formed. Sometimes these may remain small and thin, never reaching the ground (Fig. 4.8), while in other cases the elongation and secondary thickening are stronger, and when this kind have penetrated the soil they form **stilt-roots** (Figs 4.7 and 4.10). Associated with these, for example in a few palms and in *Bridelia* spp., *Macaranga barteri, Klainedoxa gabonensis* and *Commiphora fulvotomentosa*, prominent **root-spines** are formed, some of which may later develop into full stilts (Jeník and Harris 1969). In some cases a great proportion of the skeleton roots is above ground and the tree is virtually held up

Fig. 4.8 Adventitious roots on the lower trunk of *Afrosersalisia afzelii*. This type of aerial root seldom develops into thick stilts reaching the soil.

only by the stilt-roots. Though in a particular tree or even a species an explanation for the adaptive value of stilt-roots may not always be straightforward, their significance for the mechanical support of a tall and unstable tree in soft ground is evident. The weight of the tree is distributed over a larger surface ('snow-shoe' effect) and the passage of nutrients does not involve the long underground pathways that are frequently liable to oxygen depletion in poorly aerated, flooded soil ('short circuit' effect).

Similar adaptive value can be predicted for the **columnar** roots developed in several *Ficus* spp. Initially, slender free-hanging roots of these trees become anchored and form secondary tissues in which tension wood is developed (Zimmermann *et al.* 1968). This results in contraction and markedly straight aerial organs that become thick and form sizeable props for the limbs of the crown (see Frontispiece). Vegetative extension of some *Ficus* trees, notably that of *F. benghalensis*, can result in 'groves' of trees in which the trunks are actually columnar roots.

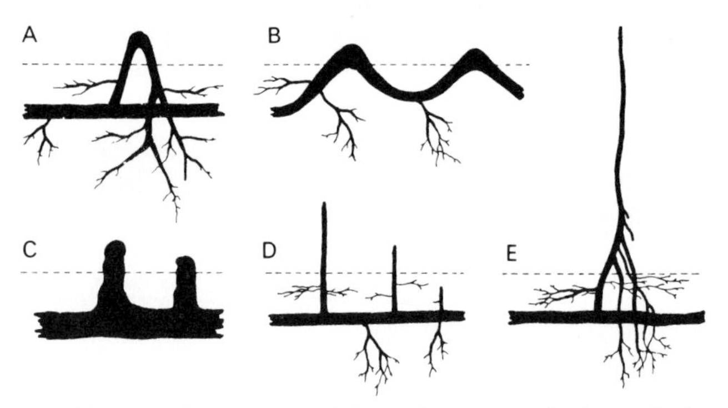

Fig. 4.9 Types of **pneumorrhizae** that occur in impeded tropical forest soils: A = lateral type of knee-root; B = serial type of knee-root; C = root-knees of *Taxodium*; D = peg-roots; E = stilted peg-root.

In impeded soils, other less conspicuous and lesser-known features of tropical roots occur, notably types of **pneumorrhizae**, which grow upwards from the soil (Fig. 4.9). *Knee-roots* consist of a loop formed where portions of roots have left the soil and then returned to it. The lateral type of knee-root develops as a lateral branch from a deeply situated horizontal mother-root, as in *Symphonia globulifera* in South America and *S. gabonensis* and *Mitragyna ciliata* in Africa. In the serial type of knee-root, found for instance in *Bruguiera* mangroves, a horizontally growing root curves repeatedly in and out of the soil. The formation of knee-roots thus depends upon the orientation of growth taken by the root apex, and involves the exposure for a certain period of its delicate primary tissues to the risk of dessication and of browsing by herbivores. Knee-roots should not be confused with the *root-knees* of the *Taxodium* type, which result from intense localised cambial activity on the upper side of a secondarily growing subterranean root.

Peg-roots are a very remarkable type of root modification which also arise from a horizontal mother-root. Unlike knee-roots they grow more or less vertically and can reach even a height of 30 cm. They are often about the thickness of a pencil, and look rather like pegs inserted into the soils (Figs 3.13 and 6.21); they are well-known in *Avicennia* mangrove swamps all over the tropical world. Peg-roots also form in freshwater swamps, for example in palms such as *Raphia* spp. and *Ancistrophyllum* spp., and in certain dicotyledonous trees, notably *Anthocleista nobilis* and *Voacanga thouarsii* (see Jeník 1971; 1978). In species of the mangrove *Sonneratia* the peg-roots show strong secondary thickening and form cone-shaped organs protruding above the ground. In *S. caseolaris* the conical pneumorrhizae can attain up to 2 m in height (Percival and Womersley 1975).

Another special type is the *stilted peg-root* which appears to be a secondary adaptation of this kind of pneumorrhiza. In *Xylopia staudtii* in tropical Africa, some of the slender peg-roots may continue growth, twisting upwards and reaching as high as 2 m. These later branch, forming a set of lateral roots, which grow down again and anchor themselves into the soil (Fig. 4.9[E]). This set of stilts occurs in addition to the ordinary adventitiously formed stilt-roots on the bole.

Many of the preceding types of aerial roots have been called breathing roots or pneumatophores, but in view of other uses of the latter word in biology, pneumorrhizae seems to be a preferable term. Their surface is commonly dotted by numerous lenticels interrupting the periderm of the secondarily grown organs in dicotyledons, or by loose powdery tissue that breaks through the cortex of the primary organs in palms and other monocotyledons. It is likely that many are adaptations to growth in soils lacking aeration, and Scholander *et al.* (1955) have shown that gaseous exchange takes place very freely in the roots of mangroves through the lenticels and large intercellular spaces. Further evidence is provided by the absence of some of these root modifications when *Mitragyna ciliata* and *Anthocleista nobilis* are growing on mesic sites. In the latter species, examples have been found where peg-roots were formed on a mesic site but only at the bottom of a man-made ditch. Another interesting point is that some pneumorrhizae, in common with many aerial roots of epiphytic orchids, contain chlorophyll and may therefore make some contribution to their carbohydrate content (see Section 6.1).

In dicotyledonous trees, four types of root systems have been tentatively classified according to the position and shape of their skeleton roots (see Coster 1932; 1933; Wilkinson 1939).

1. Root system with thick horizontal surface roots, frequently merging into large spurs or buttresses; with weak vertical sinkers and tap-root entirely absent.
2. Root system with thick horizontal surface roots, and well developed tap-root and sinkers.
3. Root system with weak surface roots, and a rich system of many oblique 'heart' roots and a prominent tap-root.
4. Root system with numerous sizeable aerial stilt-roots, and a network of weaker underground roots.

Among emergent trees, types 1 and 2 clearly prevail, while type 3 is most frequent in smaller trees and large woody climbers. Type 4 is most characteristic of trees growing in waterlogged sites, although species like the African *Uapaca guineensis* and *Xylopia staudtii* (Fig. 4.10) develop stilt-roots even on mesic soils. The size, shape and position of skeleton roots also vary according to the age of the tree. Some types of tap-root are produced in nearly all seedlings but other roots may later become predominant (see Mensah and Jeník 1968).

More recently, the diversity of roots, root subsystems and root systems in tropical trees has been summarised according to

(a) capacity of the root axes for secondary thickening;
(b) general structure of the root skeleton in adult specimens;
(c) relationships of terrestrial and aerial root subsystems;
(d) changes of roots during the life-span of a tree;
(e) occurrence of 'abnormal' root forms assumed to represent inherited genotypic adaptations.

Using these criteria Jeník (1978) described 25 organisation models, named according to common and/or well studied species of tropical trees (Fig. 4.11).

Variation in the shapes and branching habits of tropical tree trunks and crowns is much greater than in the temperate zone, and has been subject to many investigations since the publication of the stimulating book *Wayside Trees of Malaya* by Corner in 1940. A recent classification of morphogenesis and branching in tropical trees identifies 23 architectural tree models, named according to distinguished botanists and foresters (Hallé and

Fig. 4.10 Stilt-roots of *Xylopia staudtii*, growing on a mesic site in the Atewa Range, Ghana. On impeded soils, this species may also develop stilted peg-roots.

Oldeman 1970; Hallé *et al*. 1978). This system (see Fig. 4.12) is based mainly on criteria related to shoot elongation, such as the life-span of terminal apices and on the differentiation of terminal and lateral meristems. Depending on which apex is being considered, several different alternatives may affect the architecture: reproductive or vegetative differentation, plagiotropic versus orthotropic orientation, rhythmic as against continuous growth, and a variety of timings and origins of branching (see also Tomlinson 1978).

Accordingly, trees are subdivided into unbranched (monoaxial) and branched (polyaxial) categories, the latter having vegetative axes that are either all equivalent or are not. Again, the branched trees with axes that are not equivalent are divided into those

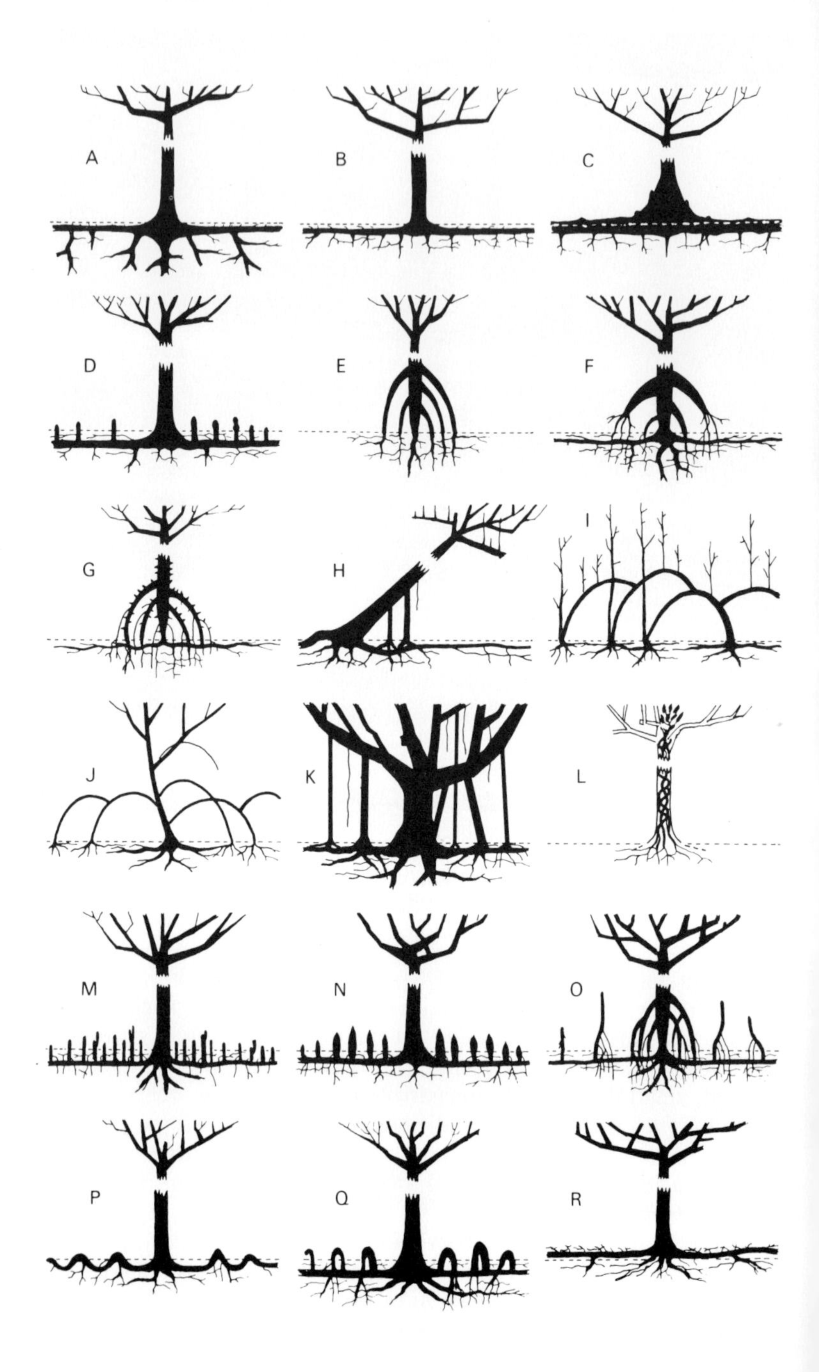
A
B
C
D
E
F
G
H
I
J
K
L
M
N
O
P
Q
R

having heterogeneous axes (differentiated into orthotropic and plagiotropic shoots), and homogeneous axes (either all orthotropic or all mixed). Yet more subdivisions are made according to sympodial or monopodial height growth, rhythmic or continuous extension, 'basitonic' or 'acrotonic' branching, short- or long-lived branches, etc.

The identification of such characteristics of primary shoot growth and the recognition of the architectural model are easiest in young trees. In a mature specimen, the shape may have been greatly modified through progressive secondary thickening, by physiological stress or because of natural disturbance to the forest. In addition, trees sometimes show 'reiteration', which means the addition to the existing plant body of a new shoot system that conforms rather closely to the architectural model which had previously been expressed by the same tree (Oldeman 1974). The crowns of some old trees can feature enormous near-horizontal limbs up to 20 m in length, that provide support for large gardens of epiphytes (Fig. 3.17). In one way or another, the architectural model tends to remain reflected in the branching habit (Fig. 4.13).

The majority of tropical trees possess a smooth, light-coloured bole of a cylindrical shape. Some, however, show features such as fluting, spiral twisting, large spines as in *Fagara macrophylla* or *Erythrina mildbraedii* (utilised as 'rubber stamps'), adventitious roots such as those in Fig. 4.8, etc. These can be very useful in identification, as can the shades of bark colour, though this can be influenced by the crustaceous lichens present. Ridged and furrowed bark or rhytidome is not nearly as common as in temperate trees, but features of the bark are very distinctive when displayed in a 'slash', cut through the outer and inner bark to the young sapwood (Fig. 4.14). Characteristics such as its colour, texture, odour and taste, and also the presence of latex

---

Fig. 4.11 Major organisation models of root systems in dicotyledonous tropical trees (A) *Chlorophora excelsa*; (B) *Cariniana pyriformis*; (C) *Piptadeniastrum africanum*; (D) *Xylocarpus mekongensis*; (E) *Uapaca guineensis*; (F) *Tarrietia utilis*; (G) *Bridelia micrantha*; (H) *Protomegabaria stapfiana*; (I) *Scaphopetalum amoenum*; (J) *Rhizophora mangle*; (K) *Ficus benjamina*; (L) *Ficus leprieuri*; (M) *Avicennia germinans*; (N) *Sonneratia alba*; (O) *Xylopia staudtii*; (P) *Bruguiera gymnorrhiza*; (Q) *Mitragyna stipulosa*; (R) *Alstonia boonei*. (From Jénik 1978.)

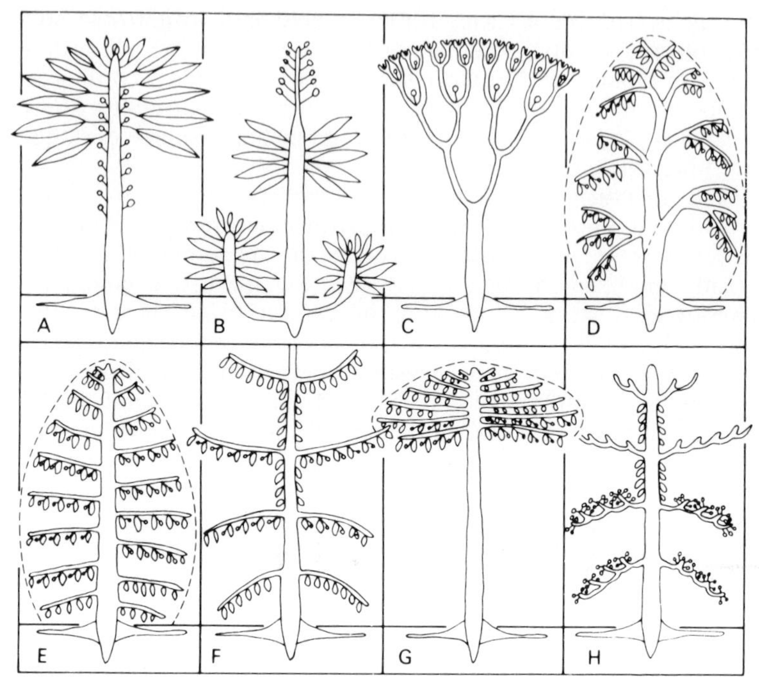

Fig. 4.12 Architectural models of shoot systems in tropical trees: examples of the classification by Hallé and Oldeman (1970):

Unbranched trees:

(A) Corner's Model, e.g. *Cocos nucifera, Elaeis guineensis, Ficus theophrastoïdes, Mauritia flexuosa, Phyllobotryum soyauxianum, Pycnocoma angustifolia.*

Branched trees:

branches all of equal status

(B) Tomlinson's Model, e.g. *Euterpe oleracea, Raphia gigantea.*

(C) Leeuwenberg's Model, e.g. *Anthocleista nobilis, Alstonia sericea, Manihot esculenta, Plumeria acutifolia, Rauvolfia vomitoria.*

(D) Troll's Model, e.g. *Cassia javanica, Delonix regia, Parinari excelsa, Piptadeniastrum africanum, Pterocarpus officinalis.*

branches of different status

(E) Rauh's Model, e.g. *Artocarpus incisa, Cecropia peltata, Entandrophragma utile, Hevea brasiliensis, Khaya ivorensis,*

Musanga cecropioides, Pentadesma butyracea, Triplochiton scleroxylon.

(F) Massart's Model, e.g. *Anisophyllea* spp., *Araucaria columnaris, Ceiba pentandra, Diospyros matherana, Pycnanthus angolensis.*

(G) Cook's Model, e.g. *Canthium glabriflorum, Glochidion laevigatum, Panda oleosa, Phyllanthus mimosoides.*

(H) Aubréville's Model ('pagoda trees'), e.g. *Manilkara bidentata, Omphalocarpum elatum, Sterculia tragacantha, Terminalia catappa, T. ivorensis.*

Fig. 4.13 Various crown shapes of tropical forest emergents. From left to right: *Tarrietia utilis, Pentadesma butyracea* and *Lophira alata*. Below, *Musanga cecropioides* can be seen colonising the margin of the stand. (W Africa, December/January.)

Fig. 4.14 Bark 'slash' on *Celtis mildbraedii*, showing the conspicuous tangential bands that assist in identification of this emergent species (the dark zones being a bright red-brown colour).

and various other exudates, can be used by experienced foresters, tree spotters and botanists for rapid recognition in the field. The anatomical differences which underlie the slash characters have for the most part not been studied scientifically (but see Whitmore, 1962*a*; *b*), but in some countries the local knowledge of barks for medicinal purposes and poisons may be very advanced. (Indiscriminate tasting of bark is not advisable, therefore.)

New branches arising from dormant epicormic buds on undamaged older trunks are relatively uncommon, except for example in swamp species such as *Protomegabaria stapfiana* and *Grewia coriacea* in tropical Africa. Vegetative reproduction by aerial roots or root suckers (see section 5.7) occurs in a few dicotyledons, such as *Scaphopetalum amoenum*, and in a number of monocotyledons, for example *Pandanus* spp. and bamboos. Vegetative spread by means of columnar roots shown by banyan and other *Ficus* spp. is among the most spectacular dispersal processes in nature (see Frontispiece).

Terminal and lateral buds on tropical tree shoots are often smaller than in temperate regions, and frequently lack clearly defined bud-scales, though these tend to be more common in flower buds. In some cases the dormant growing points may be surrounded by stipules, hairs or resinous secretions, and it has been suggested that these may hamper the movement of herbivores such as lepidopterous caterpillars, diverting their activity on to older foliage.

Some of the attractive features of young leaves are discussed in section 5.4. The final size that a leaf attains is extremely variable, lengths ranging from a few millimetres to several metres. A classification of leaf sizes is given in Table 4.4 from the leptophyll category (< 0.25 cm^2) to the megaphylls exceeding 16.4 dm^2. An altitudinal progression was found in this study of leaf size in relation to climate at 38 sampling sites in Costa Rica (Dolph and Dilcher 1980). The percentage of species having large leaves (notophyll and above) averaged 82 per cent in lowland and premontane forests, 46 per cent in lower montane forests and 5 per cent in montane forests. The predominant size in a lowland Ghanaian forest was 20–45 cm^2 (notophyll) although larger leaves were more frequent in wetter sites (Hall and Swaine 1976).

Table 4.4 Size classes of leaves in the trees, shrubs and vines found at three sampling sites at different altitudes in Costa Rica. (After Dolph and Dilcher 1980)

Leaf area (cm^2)	Leaf size class	% Distribution of size classes		
		SITE A Finca Las Cruces (premontane)	SITE B Near La Chonta (lower montane)	SITE C Summit of Cerro de la Muetta (montane)
	leptophyll	—	—	14
0.25 —	nanophyll	—	—	57
2.25 —	microphyll	—	50	29
20.25 —	notophyll	22	36	—
45 —	mesophyll	67	14	—
182 —	macrophyll	11	—	—
1640 —	megaphyll	—	—	—
Number of species		36	34	7
% with large leaves (notophyll and above)		100	50	0

Mature leaves vary considerably, although the visitor to the tropical forest is often struck by the dominance of dark green, oblong-lanceolate, entire and leathery leaves (or leaflets) which make the foliage within the forest appear very uniform. Roth

(1984) explains the leathery leaf structure in tropical trees as a product of water stress acting during even short dry periods that affect the upper canopy of rain forests as well as evergreen seasonal forests (see section 3.3). Originally, the transformation of leaf structure may have been induced by the actual impact of microclimate, but during the progress of evolution it evidently often became a positively selected feature. A striking feature of the majority of leaves in the undergrowth layer is the presence of drip-tips, mentioned in section 3.3. The quicker drying they appear to cause will have the effect of enhancing transpiration, and it is possible that it also reduces the tendency for epiphyllous plants to colonise and mask the leaf surface (see Fig. 4.24).

Emergent trees with their crowns exposed to strong insolation and intermittently drying conditions generally have much smaller leaves without a distinct acumen (Fig. 5.20). Some of these are capable of various movements with regard to the sun's position and the time of day (see section 5.4). The position of the leaflets in many leguminous trees, for example *Piptadeniastrum africanum*, alters so much that around midday the light intensity in the lower stories is substantially increased. These movements are brought about in many tropical leaves by specialised pulvini or leaf-joints (see Fig. 5.10). Another feature is the tendency of riverside species to develop linear-lanceolate leaves.

The patterns of leaf-shedding and flushing of new leaves will be discussed later in Chapter 5. These changes in the canopy are obviously of basic importance to the life of the ecosystem, and help to determine the general physiognomy and classification of tropical forests (see section 6.5).

The diversity of flowers, fruits and seeds is remarkable, with a great variety of generative organs and many pollination and dispersal mechanisms. Detailed description of flower and fruit morphology, so fundamental in the classification of plants, is outside the scope of this book, but may be obtained from local floras. Some experimental details on flowering and fruiting will be described in sections 5.9 and 5.10, while a few important characteristics are mentioned here.

Hall and Swaine (1981), applying an earlier classification (Faegri and van der Pijl 1979), identified eight blossom types among the great variety encountered in West African forests: funnel (e.g. *Rothmannia longiflora*), dish (*Mussaenda chippii*), bowl (*Monocyclanthus vignei*), bell (*Cola reticulata*), trap (*Ceropegia gemmifera*), brush (*Mimosa pudica*), gullet (*Leonotis nepetifolia*) and flag (*Berlinia tomentella*).

An interesting feature is the occurrence of flowers and inflorescences (and consequently fruits) on leafless woody trunks and

Fig. 4.15 Fruits borne directly on the main stem, following cauliflory (in the narrow sense – see Table 4.5). In this pigmy tree, *Allexis cauliflora*, the single seed is projected explosively from the capsule.

larger branches, which is known as 'cauliflory' (Fig. 4.15). This phenomenon in its broad sense is subdivided in Table 4.5 and discussed further by Richards (1952). A very striking case is the positioning of the flowers on the leaves, which we suggest may be termed 'phylloflory'. This is shown in several genera of Flacour-

Table 4.5 Cauliflory (broad sense) and its subdivision

Cauliflory	Flowers on the trunk, leafless twigs and roots
Ramiflory	Flowers on larger branches and leafless twigs, but absent from the trunk
Trunciflory	Flowers on the trunk, but not on the branches and twigs (equals cauliflory in the narrow sense)
Basiflory	Flowers at the base of the trunk
Flagelliflory	Flowers on pendulous twigs spreading down from the lower part of the trunk on to the ground surface
Rhizoflory	Flowers on the roots

tiaceae, as for example in the pigmy tree *Phyllobotryum soyauxianum* (Hutchinson *et al.* 1954) where the inflorescences are borne on the midrib of the assimilating leaves. In Papua New Guinea, *Chisocheton pohli* and *C. tenuis* (Meliaceae) flower on the rachis, between periodic flushes of leaflets produced by an indeterminate leaf-tip meristem.

Cauliflory and phylloflory, like other adaptations in floral biology, can best be understood in terms of the predominant pollination agencies. Within the forest the air movement is generally very low, and it is usually found that insects, and also birds and bats play important roles in the transport of pollen. For example, ants are constantly passing along the trunks (Fig. 4.26), branches and leaves of many trees and can easily serve as a vehicle for pollen.

With regard to fruit and seed dispersal, a large proportion of trees in tropical forest also appear to be adapted to animal activity. Even in the upper tree layer only about half the species are dispersed by wind, the majority of the remainder being distributed by birds, bats, rodents, monkeys and insects. In the middle and lower tree storeys animal agents prevail (Jones 1956). Observations in tropical forests of Panama and Costa Rica (Leigh and Windsor 1982) showed the dominating role of birds in dispersal of fruits of both overstorey and understorey species (Table 4.6). In the West African forests Hall and Swaine (1981) found that about 70 per cent of the tall trees and climbers had fleshy fruits in nearly all the forest types studied.

Table 4.6 Fruit dispersal in two different tropical forests. (After Leigh and Windsor 1982)

Means of dispersal	Number of species dispersed at			
	Barro Colorado, Panama		La Selva, Costa Rica	
	Overstorey	Understorey	Overstorey	Understorey
Birds	28	58	42	63
Bats	10	10	6	5
Primates	7	2	8	2
Terrestrial mammals	4	6	9	5
Elephants	0	0	0	0
Wind	24	0	12	1
Mechanical (explosive)	1	6	2	3
Unknown	3	6	14	14
Total	77	88	93	93

Some fruits are heavy – for example, those of *Omphalocarpum* spp. – and perhaps are therefore more likely to penetrate the canopy and reach the soil surface (Fig. 5.30). Some trees have explosive fruits dispersing the seeds over large areas around the mother tree. In *Hura crepitans*, a large thorny tree native to tropical America, the dehiscent capsule can eject the seed more than 40 m (Swaine and Beer 1977). Like many members of the Violaceae, *Allexis cauliflora* (Fig. 4.15) lobs its single seed into the undergrowth of the African forests (Jeník and Enti 1969). Seeds and fruits of wind-dispersed trees often have wings or plumes, and some can drift or glide in virtually still air.

4.3 Other plants

In the previous section various woody plants have been described whose growth habits are rather far from the common impression of a 'tree', exemplifying the problems of definition. Many of the smaller woody species remain shorter than the arbitrary 5 m minimum height, some branch from the base or develop a cluster of vertical main stems, while others may form a single trunk but lack solid woody tissue, and so on. Conversely, certain herbaceous plants of the tropical forest can achieve a size and even a structure comparable with conventional trees. The palms are a good example of a family that contains such a wide spectrum of life-forms that it would cause insurmountable difficulties to try and constrain it to fit into Raunkiaer's classification of plants.

Of the apporoximately 2 770 species of **palms**, many grow scattered through the closed-canopy, broadleaved tropical forest, although some tend to create monospecific stands on waterlogged sites. The diversity of growth forms, and considerable reproductive capacity and competitive ability of palms (Corner 1966; Dransfield 1978) contribute towards their r-strategy. Besides tree forms, the shrub palms, acaulescent palms and climbing palms are often able to occupy temporarily open sites such as forest clearings, exposed river banks or areas that have been burnt. In this connection, Hallé *el al.* (1978) point out that a single specimen of *Corypha elata*, a palm found in India and Sri Lanka, can at the end of its life produce some 3–15 million functional flowers and 300 000 fruits.

Bamboos occur quite frequently, especially in tropical uplands where there are wet depressions and slopes, and they spread readily by rhizomes. In the Andes, large patches in the montane and sub-alpine forests are covered with thickets of the genera *Chusquea* and *Guadua* (see Fig. 6.19), whose taxonomic classification is still far from complete, since their vegetatively spreading clones remain sterile for several decades. The Asian *Dendrocalamus giganteus* and *Gigantochloa aspera* can reach heights of up to 30 m and are well known for their 'intolerance' towards other plant members of the forest community.

Fig. 4.16 Tree-ferns belonging to an unknown *Cyathea* sp. growing in montane forest in E Ecuador.

Inseparable from the general picture of tropical forests are **tree-ferns** of the Cyatheaceae (Fig. 4.16). Some are light-demanding species that grow along river banks and in larger clearings; there are also shade-bearing types which survive in dark ravines or in the dim phase under the closed canopy of mature forest.

One of the characteristic features of all kinds of tropical forests, including mangrove stands, is the presence of numerous **epiphytes**. Tixier (1966) divided this important component of the vertical structure into macro-epiphytes (vascular plants) and micro-epiphytes (mosses, liverworts, algae and lichens). In a listing of the former, Madison (1977) includes the following families containing at least 400 species: Aspleniaceae (400 species), Hymenophyllaceae (500), Polypodiaceae (970), Araceae (850), Bromeliaceae (919), Orchidaceae (20 000), Ericaceae (483), Gesneriaceae (549), Melastomaceae (483), Moraceae (521) and Piperaceae (710). Macro-epiphytes are particularly numerous in montane and moist forests (see Figs 4.17, 4.35 and 1.1). Micro-epiphytes are also very abundant, although less impressive in terms of their species diversity. In a mountain district of Vietnam, there were 116 species of mosses, 110 of liverworts and 36 of lichens (Tixier 1966).

Fig. 4.17 Base of a tropical forest tree (zone 5 in Fig. 4.18), richly covered with epiphytic mosses, filmy ferns and *Calvoa monticola* (Melastomaceae).

The substrate for all these epiphytes may be the rough bark surface, crevices within the rhytidome, moist crown humus lying over large limbs, pockets of soil, or deserted ant and termite nests. Of special interest are the epiphyllous species: liverworts, lichens or algae growing on the surface of living leaves (see Fig. 4.24).

Epiphytes exist under very diverse conditions according to their exact position in the forest structure, and a classification of these micro-habitats is given in Fig. 4.18. Sudden transitions can occur, for example if the zone 2 epiphytes, which normally experience semi-arid conditions with full radiation (Fig. 3.17), fall down into the moist, shady undergrowth on a broken limb or uprooted tree. Such an abrupt ecological stress has few parallels in other plant communities, and it may have been a factor in the evolution of rich tropical floras (section 4.5).

Climbers of various sizes, morphology and life-form add to the complexity of the vertical structure. The woody climbers or lianes can become very large, with thick convoluted stems (Fig. 4.19).

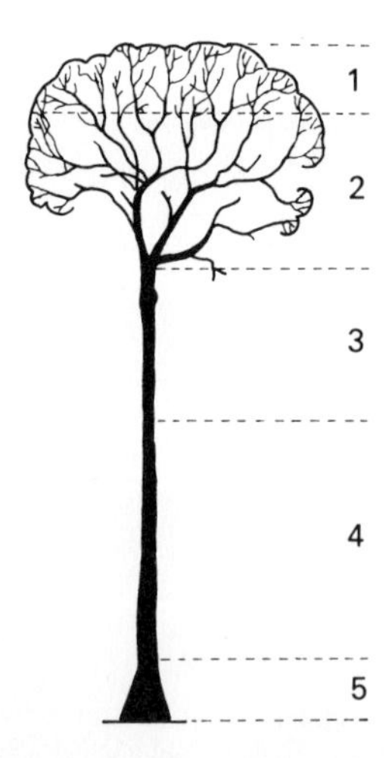

Fig. 4.18 Schematic distribution of microhabitats for epiphytes on an emergent tree in the tropical forest: (1) fully exposed apical portion of the crown with microepiphytes; (2) main zone of epiphytes covering large limbs; (3) drier upper part of bole with crustaceous lichens; (4) moister lower part of the trunk with lichens and frequent bryophytes; (5) base of the tree, covered with bryophytes, particularly between any buttresses and spurs.

Fig. 4.19 Base of stem of a large liane, *Neuropeltis prevosteoides*, showing contortions due to loop formation and abnormal secondary thickening (see Fig. 5.15).

These can show as many as seven different types of abnormal cambial activity (Obaton 1960), which result in an irregular outline of the stem (Fig. 5.15), and even occasionally in a rope-like appearance due to the partial disintegration of the tissues (see also Fig. 1.3).

The smaller climbers are more tolerant of the shade in the forest interior and include many herbaceous species (vines), such as some of the Araceae. Both woody and herbaceous climbers can be classified according to their organs of attachment to the supporting tree: for example, stranglers, twining climbers, root-climbers, and tendril-climbers. One can also distinguish between climbers germinating in the soil and those which start as epiphytes in the crowns and send down roots which eventually reach the soil. In the latter group is the remarkable life-form of the strangler (Figs 4.20, 4.21 and Frontispiece). The descending roots thicken, branch and anastomose, ultimately forming a compact casing around the 'host' tree, which is then overgrown and literally strangled, leaving the former epiphyte as an independent tree on its own roots. Stranglers are numerous in the genera *Ficus*,

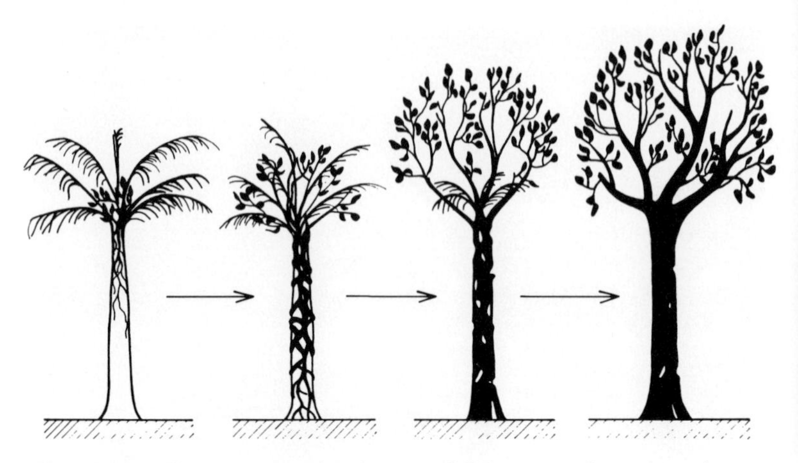

Fig. 4.20 Four stages in the establishment of a strangler – *Ficus leprieuri* on *Elaeis guineensis.*

Schefflera and *Clusia*; sometimes they may even become members of the upper tree layer.

Ferns and fern-allies have a strong representation both among the epiphytes and in the **herb** layer of most tropical forests. Families such as the Lycopodiaceae, Selaginellaceae, Hymenophyllaceae, Adiantaceae, Thelypteridaceae and Aspidiaceae are represented in all tropical regions. Some ecologists consider the ratio of ferns to angiospermous herbs as a distinctive characteristic of particular types of tropical forest communities.

Dicotyledonous families commonly represented in the herb layer include, for example, the Rubiaceae, Gesneriaceae, Begoniaceae, Melastomaceae and Acanthaceae; while monocotyledons often belong to the Cyperaceae, Gramineae, Commelinaceae, Marantaceae, Zingiberaceae and Araceae (see Figs 4.22 and 4.23). A number of the ground herbs show interesting features of shape, colouring and surface structure of leaves which have made some of them popular house plants, – for example, species of *Begonia, Saintpaulia, Calathaea, Anthurium*, etc.)

Tropical forests are well known for their richness in parasitic flowering plants. Mistletoes (family Loranthaceae) are typical examples of this life-form in tropical forests, with about 700 representatives described so far. Possessing green leaves, the mistletoes produce some of their own carbohydrates, and thus belong to the group of hemiparasites. Their frequency in tropical tree crowns is best demonstrated by the prevalence of what has been called hyperparasitism (Kuijt 1969), in which the haustoria

Fig. 4.21 Well-established strangling fig. Note the anastomosing roots, including one that is clearly inhibiting cambial activity in the host tree, which is now nearly encased in the 'trunk' (roots) of the strangler.

Fig. 4.22 Characteristically undulating leaf surface of *Psychotria ankasensis* (Rubiaceae), a herb found in the undergrowth of W African forests.

of one species become established in the stems of another mistletoe species. The Santalaceae is another family containing a number of root and stem parasites, including the sandalwood (*Santalum album*), which has a valuable, scented wood and oil. However, the most remarkable of the tropical parasites are probably the Rafflesiaceae, particularly the puzzling *Rafflesia arnoldi*, which has a flower about 1 m in diameter, and an almost complete reduction of the vegetative organs that are normally present in vascular plants.

Much less spectacular are the vascular holo-saprophytes – mostly small plants scattered in the dim field layer. They belong to the Gentianaceae (*Voyria*, *Leiphaimos*, *Sebaea*), Burmanniaceae (*Burmannia*, *Gymnosiphon*) and even Orchidaceae (*Auxopus*). According to Schnell (1970–71) most of these saprophytes possess extremely light seeds, which enables them to be dispersed by means of the slightest air movements occurring near the ground in a closed forest.

Mosses and liverworts form an omnipresent component of the epiphytic gardens and epiphyllous flora (Fig. 4.24). Shade-bearing species cover the lower part of tree trunks (for example

Fig. 4.23 The broad leaves and inflorescence of *Mapania coriandrum*, a tropical forest sedge.

Fig. 4.24 Epiphyllous flora on the surface of a living leaf, consisting of lichens, mosses, liverworts and algae.

Frullania, Lejeuneaceae), while the few drought-resistant species make up a part of the crown gardens (for instance *Macromitrium* spp.). In mist forest, festoons of bryophytes such as *Meteriopsis* and *Squamidium* often hang down from branches. Green and blue-green algae may be present in very large numbers on the surfaces of rocks, soil and bark, and as part of the epiphyllous flora on leaves (Fig. 4.24). They have been little studied, although some of them may contribute to nitrogen fixation in the ecosystem (see section 3.3). The same is true of certain genera of bacteria and actinomycetes, and although these and the fungi (Fig. 6.5) and lichens (Fig. 3.3) are not classified as plants, some of them form close associations with trees (see section 6.3), while others play an important role in decomposition (section 6.2) and parasitism.

4.4 Animal life

Such is the diversity of animal species in the tropical forests that it will only be possible to give the briefest hint of it here, and to describe a few of the most striking interactions between animals and plants in section 6.3. Fortunately in recent years a number of excellent television documentaries and films have been made, with high quality wildlife photography that shows fauna in its natural habitat. An increasing number of symposium volumes and books are also being produced which collate the available information on the different groups (for example, UNESCO 1978; Sutton *et al.* 1983; Golley 1983a).

Invertebrates are by far the most numerous, both by species and individuals present. Many play a vital role in the ecosystem, for example as the decomposers responsible for the turnover of materials and energy (see section 6.2). Notable amongst them are many microscopic groups – a study of micro-arthropods, for instance, indicated numbers of about 38 000 per m^2 in the surface soil of tropical forests in Nigeria (Madge 1965). Large land-snails and millipedes are a feature of some areas, while earthworms that form large and frequent above-ground casts (Fig. 7.10 B) mix the soil up, incorporate organic matter and may improve the exchange capacity and content of some nutrients.

Arthropods are found everywhere in the forest, usually actively carrying out the functions of herbivore, carnivore (Fig. 4.25) or decomposer; and themselves representing a major food resource for insectivorous vertebrates such as lizards, birds and ant-eaters. They also introduce an additional level of organisation to the ecosystem through the highly co-ordinated social behav-

Fig. 4.25 Spider's web, made by *Argiope* sp., showing white 'stabilimentum' in the centre (Atewa Range, Ghana).

iour shown by some of the colonial taxa. Perhaps the most frequently seen of these are the ants, running up and down the trunks (Fig. 4.26), branches, foliage and fruit; across the forest floor and into crevices in the soil; or travelling in 'drives' with soldier-ants as guards, and perhaps even returning with another species of ant as slaves. They make nests, some of them quite large, partly or wholly underground, on tree boles or constructed by drawing leaves together and joining them.

Permanent structures rising as much as 1.5 m above-ground are built by *Atta* spp, with the openings sometimes protected by 'chimneys' 10 cm tall, and the nest extending horizontally even to 250 m^2 (Cherrett 1983). This is a genus of leaf-cutting ants, and the several million individuals in a colony have been estimated to collect at a rate exceeding 60 g/h dry weight, carrying the pieces to specialised fungus-gardens where they are chewed and digested by the combined proteolytic enzymes of the fungus, the leaf, and the ant's faecal liquid (UNESCO 1978: p. 161). They have been estimated to harvest approximately 0.2 per cent of the gross primary production of the forest, also reducing it by about 1 per cent through defoliation. However, they seldom severely damage their resources, perhaps because of their complex foraging strategy which results in nearby trees often being ignored in

Fig. 4.26 Ants running up and down the trunk of *Anisophyllea griffithii* in a Malaysian forest.

favour of those of the same preferred species that are more distant, exceptionally to around 100 m. On the other hand, in pure *Citrus* plantations, they can become a serious pest, killing many trees by repeated defoliation.

Fig. 4.27 Termites crossing the floor of the forest at Semongoh in Sarawak, carrying pellets containing the inoculum for their fungal gardens.

Some termites (Fig. 4.27) also cultivate fungal strains (including those of the basidiomycete *Termitomyces*), and build large, conspicuous mounds on the forest floor, or smaller nests in dead wood, up in trees or underground. They are primarily decomposers, being responsible for consuming between 1 and 16 per cent of the total litterfall in Malaysian forests (Collins 1983), and apparently even more in Zaire. Flies, bees, wasps and beetles were the most numerous flying insects in a study from mid-February to mid-March at four heights in a forest in Sulawesi, an average of more than 1 500 being caught at each trapping period (Sutton 1983). The majority were to be found at the highest traps (26–30 m), and there was also pronounced horizontal diversity found in the forest.

Amongst vertebrates as well as invertebrates there are some examples of mimicry with the background environment, as for example in the reptiles (Figs 4.28 and 4.29) and amphibia (Fig. 4.30). The total density of birds in tropical forests is often roughly comparable to those in temperate forests in the summer, but there are many more tropical species, particularly in South America, including a number that are quite rare (UNESCO 1978: p. 143). Above about 1 000 m altitude, however, species richness declines sharply. Particularly in rain forest, the breeding season is

Fig. 4.28 Mimicry with its background shown by *Bitis gabonica*, the Gaboon viper. Note especially the markings on the old, sloughed-off skin.

usually longer than in the temperate zone, and clutch sizes about half as large. The birds often restrict themselves to quite a limited vertical and horizontal range, and tend to lose young more through predation.

Mammals such as rodents appear to be about as common in tropical as in temperate forest, and to have a more extended breeding season with longer intervals between litters. Mammals are not so frequently seen as might be imagined from documentary films, often being rare, shy, nocturnal or limited to a particular layer (Fig. 4.31). Population numbers can be relatively low, especially in large herbivores that require a large tract of forest to feed in, and in carnivores. Exceptions to this are some of the primate species in South-East Asia and Africa, where there may be a greater chance of seeing groups of them, particularly if the localities are known in which they feed or rest at particular times of day. Another striking instance is that of bats, which can sometimes be seen at dusk, flying in flocks containing very large numbers of individuals. These remarkable mammals have a number of close relationships with plants, notably in pollination and fruit dispersal (Ayensu 1974); other interactions of this kind will be described in section 6.3.

Fig. 4.29 The common hinged tortoise, *Kinixys erosa*, relatively inconspicuous on the floor of a virgin forest.

Fig. 4.30 The toad *Bufo superciliaris* resembling a dead leaf on the floor of a W African forest.

Fig. 4.31 Distribution of mammals (excluding bats) within the vertical structure of the lowland rain forests of Sabah, according to the timing of their active period. (After MacKinnon 1972.)

4.5 Why so species-rich?

As already mentioned in the preceding sections, one of the remarkable features of tropical forests is their structural diversity and particularly the richness in species and life-forms that many of them contain. If stands on very dry or waterlogged sites are excluded, together with areas subject to other physiological stress and also young secondary stages, mature forests typically contain many hundreds of different kinds of plants and animals. Indeed, they are perhaps rivalled only by some coral reef communities in the numbers and variety of taxa present, and together probably account for almost half of all the Earth's gene pool.

It should also be remembered that the total number of plant species given in many local floras is still far from complete. For example, the monographer of the Chrysobalanaceae, an important neotropical family of trees and shrubs, has himself described 20 new species in the five years since he completed the monograph (Prance 1979). Epiphytes and climbers, too, are rather poorly recorded and it has been estimated that many hundreds of species remain to be described, in the case of orchids and bromeliads in South and Central America, and orchids in New Guinea. In the animal kingdom, the number of unnamed insects alone must be enormous, particularly the beetles.

On the other hand, certain biogeographical provinces, such as West Tropical Africa, which already has a second edition of its flora (Hutchinson *et al.* 1954–72), are better documented. Here it is possible to identify the majority of vascular plants, particularly the phanerophytes which provide the bulk of the forest structure. In five types of tropical forest in Zaire, for instance, 90–100 per cent of all plants were phanerophytes, the few others being geophytes (4–10 per cent), chamaephytes (2–4 per cent) and hydrophytes (2–3 per cent). There were no representatives of the hemicryptophytes and therophytes (Évrard, 1968).

In South-East Asia it is usual to find more than a hundred different species of trees per hectare, excluding seedlings (Wyatt-Smith 1953), while some recent estimates suggest that occasionally the total number of woody species may be almost 400 per hectare. The relatively poorest tropical forest region is the African, where less than 100 woody species per hectare is typical, but this is still much greater than most other types of forest. Even just to enumerate the trees present over a whole hectare is very laborious, and if the other vascular plants or indeed the bryophytes are to be included the task becomes even greater. Yet because the horizontal structure is so patchy the minimum area which is theoretically desirable is about 4 ha.

Samples of this size, however, are hardly manageable in the course of normal ecological studies, and one generally has to rely on statistically questionable extrapolations from smaller enumerations. Figure 4.32 gives a selection of species/area curves for trees in various parts of the tropics, indicating how unreliable a small sample plot can be; it also shows the especial richness of the Indo-Malaysian forests. A rare attempt to compare tree diversity with that of all vascular plants is shown in Fig. 4.33, based on the work of Lawson *et al.* (1970).

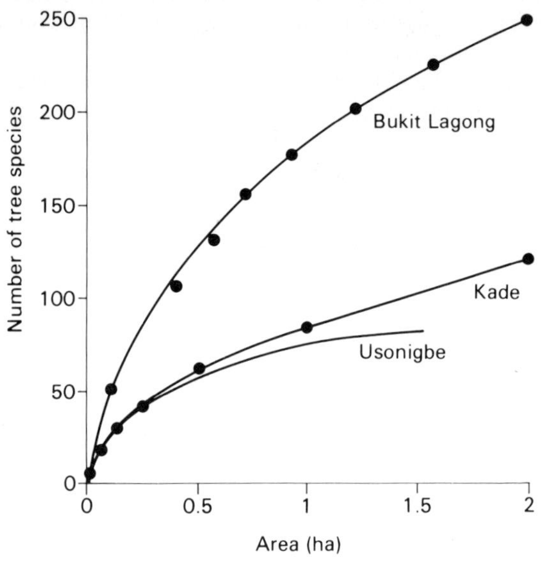

Fig. 4.32 Species/area curves for tree species in tropical forests: Bukit Lagong, W Malaysia (data from Forestry Research Institute, Kepong); Kade, Ghana; Usonigbe, Nigeria. (From Hall and Swaine 1981.)

Some of the important families and genera found in different tropical forest regions are given in Table 4.7 from which it can be seen that some are pantropical while other taxa occur mainly or entirely in one part of the world. American, African and Malayan tropical forests are often characterised by the frequency of leguminous trees (Caesalpiniaceae, Mimosaceae, Papilionaceae) which commonly appear as emergents in the upper tree layer. The same synusia is often dominated by dipterocarps (Dipterocarpaceae) in the Indo-Malaysian region, where the presence of families such as the Fagaceae, which have tree representatives in temperate forests, is worthy of note.

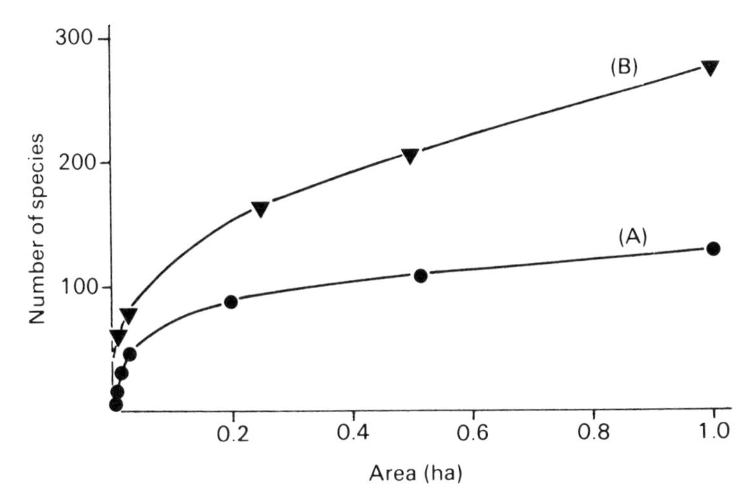

Fig. 4.33 Comparison between species/area curves for: (A) tree species; and (B) all vascular plant species, in an African evergreen seasonal forest. (After Lawson *et al.* 1970.)

At the level of genera and species, only a few taxa are naturally pantropical. For instance, *Ceiba pentrandra* is nowadays found widely in Africa, America and South-East Asia, but is only certainly indigenous to the first region (see Baker 1965). In many other cases, further taxonomical and phytogeographic studies are still needed to establish the status of closely related species, as for instance *Symphonia gabonensis* in Africa and *S. globulifera* in America (for further examples see Schnell 1961; Meggers *et al.* 1973).

In single-dominant forest and other edaphic climaxes the floristic differences between regions are more conspicuous still. In the American region, the leguminous trees *Eperua falcata* and *Mora* spp. can for instance form a high proportion of the tree layers. In Africa *Gilbertiodendron dewevrei* predominates in certain areas, while in South-East Asia and Australasia *Agathis* spp. are sometimes widespread in single-dominant stands.

The distinction between regions is also reflected in the floristic composition of the epiphytes, and the main groups can be summarised as follows (Tixier 1966):

Africa: Orchidaceae, ferns.
America: Orchidaceae, ferns, aroids, Bromeliaceae, Cactaceae.
Indo-Malaysia: Orchidaceae, ferns, Asclepiadaceae, Rubiaceae.

Table 4.7 Examples of families and genera containing dominant, abundant or subendemic species of woody plants in the chief Tropical Forest Regions

Tropical Forest Region	Family	Genus
American	Leguminosae	*Andira, Apuleia, Dalbergia, Hymenolobium, Mora*
	Sapotaceae	*Manilkara, Pradosia*
	Meliaceae	*Cedrela, Swietenia*
	Euphorbiaceae	*Hevea*
	Myristicaceae	*Virola*
	Moraceae	*Cecropia*
	Lecythidaceae	*Bertholletia*
African	Leguminosae	*Albizia, Brachystegia, Cynometra, Dialium, Erythrophleum, Gilbertiodendron*
	Sterculiaceae	*Cola, Nesogordonia, Tarrietia, Triplochiton*
	Meliaceae	*Carapa, Entandrophragma, Khaya, Trichilia*
	Euphorbiaceae	*Drypetes, Macaranga, Ricinodendron, Uapaca*
	Moraceae	*Antiaris, Chlorophora, Ficus, Musanga*
	Sapotaceae	*Afrosersalisia, Chrysophyllum*
	Ulmaceae	*Celtis*
Indo-Malaysian	Dipterocarpaceae	*Dipterocarpus, Dryobalanops, Hopea, Shorea, Parashorea*
	Moraceae	*Artocarpus, Ficus*
	Anacardiaceae	*Mangifera*
	Actinidiaceae	*Actinidia*
	Daphniphyllaceae	*Daphniphyllum*
	Dilleniaceae	*Dillenia*
	Gonystylaceae	*Gonystylus*
Australasian	Myrtaceae	*Eucalyptus, Agonis, Baeckea, Backhousia, Osbornia*
	Dipterocarpaceae	*Dipterocarpus*
	Casuarinaceae	*Casuarina, Gymnostoma*
	Himantandraceae	*Galbulimima*
	Corynocarpaceae	*Corynocarpus*
	Dilleniaceae	*Hibbertia*
	Menispermaceae	*Carronia*
	Cunoniaceae	*Ceratopetalum*

Floristic differences also appear in the composition of secondary forest. Some of the representative genera and species are:

Africa: *Harungana madagascariensis*, *Macaranga* spp., *Musanga cecropioides, Trema guineensis*.

America: *Cecropia* spp., *Miconia* spp., *Ochroma* spp., *Vismia guianensis*.

Indo-Malaysia: *Elaeocarpus* spp., *Glochidion* spp., *Macaranga* spp., *Mallotus* spp.

Considerable changes in species composition have occurred due to the activities of man. Some of the effects of widespread farming have already been mentioned, and another potent influence on local floras has been the selective removal of valuable timber species such as the mahoganies, for example *Swietenia macrophylla*, *Khaya ivorensis* and *Entandrophragma* spp. This has led to what might be termed 'depleted forests', in which younger as well as older specimens of the logged species are often scarce. A more far-reaching change in local floras has occurred with the establishment of plantations of exotic species such as rubber, oil-palm (Fig. 7.15), teak, *Eucalyptus* spp. and *Pinus* spp., unless these are grown as small blocks within the forest. The significance of such trends will be further discussed in Chapter 7.

In examining the numbers of species and general biotic diversity of unmodified forests it is necessary to admit that very little is known about the potential maximum richness that could be achieved in the biosphere under present evolutionary conditions. Only in a comparative sense can one appreciate the diversity between the temperate and tropical regions or along a latitudinal gradient.

In many groups of organisms, such as the vascular plants, vertebrates and beetles, the increasing numbers of species with decreasing degrees of latitude is a biogeographical reality (Rejmánek 1976). However, there are groups of life-forms, such as algae, fungi and lichens, and certain orders of insects, that show no spectacular enrichment of species, even within the humid tropics. Janzen (1976*b*), for example, states that species abundance of parasitic Hymenoptera declines towards the Equator, possibly due to the partitioning of the harvestable insect biomass into ever smaller heterogeneous portions that cannot support as many parasites as can an average extra-tropical prey-population. On the other hand, swallowtail butterflies (Papilionidae) show many more species at low latitudes (Fig. 4.34).

Species-richness is never homogeneously distributed over the large tropical regions. Humidity is one of the important factors, but not in the seemingly straightforward sense that the more

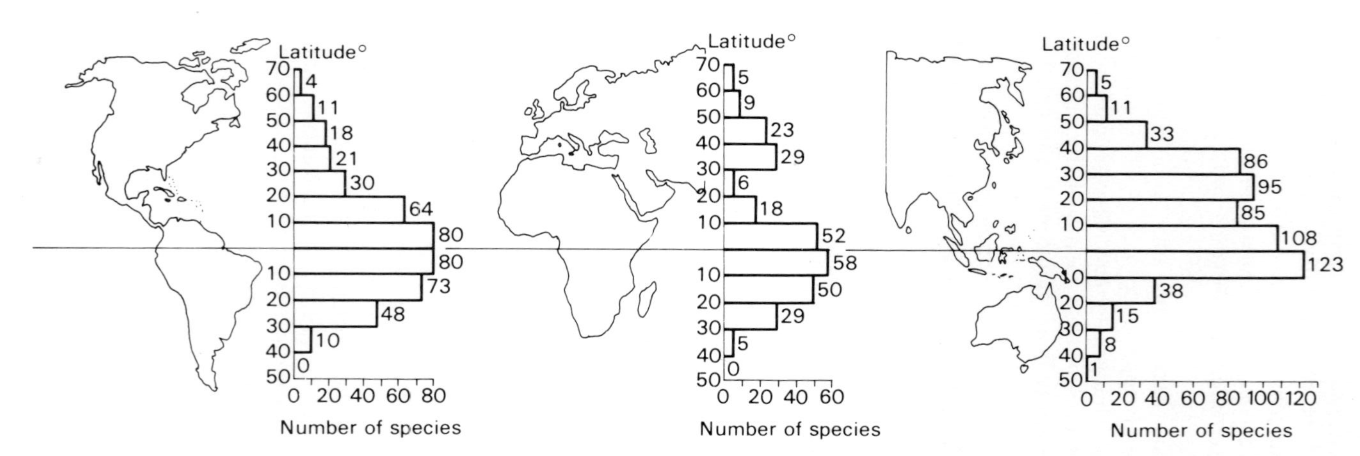

Fig. 4.34 Latitudinal gradients in the species richness of swallowtail butterflies (Papilionidae). (From Collins and Morris 1985.)

humid the environment, the more species-rich. The middle part of this geographical gradient, be it the seasonal or semideciduous forest, or even (for certain organisms), the savanna, often maintains a much richer gene pool than very humid districts such as the Amazonian *caatingas*. With regard to altitudinal gradients, many observations suggest that the middle elevations are often the richest ones (Janzen 1976*b*). Soil catenas and topographic factors also affect the representation of species in a well-known fashion. As a general rule, the maximum diversity is usually to be found in the middle part of any environmental gradient, whether it concerns moisture, temperature or nutrient substances that are toxic in excess.

In order to explain the differences in species-richness and diversity between low and high latitudes, three kinds of factors have to be considered:

1. the rate of *in situ* speciation;
2. the opportunities for *in situ* coexistence and preservation; and
3. accessibility to immigration.

All these factors combine and reinforce each other in evolution, and the formation of a new species at a particular site and time must never be taken to be the only way the ecosystem can be enriched. Speciation can also be the product of earlier *ex situ* gene flows and subsequent immigration and conservation.

New species arise, broadly speaking, from the reproductive isolation of sympatric populations (Mayr 1982 and Fig. 5.28). The factors triggering reproductive isolation can be mutation or recombination of genes, resulting either from internal genetic changes or induced by the external physical environment. The question is, are there any basic differences in environmental factors that might induce a higher rate of mutation in the tropics? The answer is – yes, since the levels of UV-B irradiation, for example, which can modify DNA, markedly increase towards lower latitudes (Caldwell 1981). As this can promote the frequency of mutations, the rate of tropical speciation might consequently be greater, and it could be further enhanced by the spatial and temporal isolation possible within the complex and highly dynamic structure of the forest. Relatively simple advantages in speciation rate can open up a broad array of ecological niches whose increasing diversity will further step up natural selection and the progress of evolution.

Newly formed species must coexist or become extinct, due to lack of space, inaccessible physical resources or biotic competition (Fig. 4.35). With regard to sufficient space, mature

Fig. 4.35 A large and diverse collection of epiphytes on a branch of *Buddleia incana* in a sub-alpine forest near the Equator in Ecuador. At least 20 species of vascular plants and bryophytes are present.

tropical forest comprises an impressive above-ground layer, yet there is little difference in available room for species in comparison with temperate forests. As regards physical resources, too, the yearly supply of solar energy in the intertropical belt is greater than at higher latitudes (Budyko 1984). If coupled favourably with other conditions such as moisture and nutrients, this may be expected to support a larger biomass, creating the potential for even more species-rich communities (see Fig. 7.1).

This potential can be realised only if biotic relationships are controlled in the sense of 'lack of dominance' and/or a 'tendency for diversity'. Hubbell (1979) assumes that species are in competitive equipoise, and that diversity expresses the balance between speciation and extinction due to chance fluctuations in numbers. Biotic factors are the main reason for the remarkable space, time and resource partitioning of the forest (Bourlière 1983). Trees themselves, by their growth form and life strategies, enable all other species to coexist because they occupy different habitats in the forest. According to Leigh (1982), tree species coexist because their reproductive rates respond differently to environmental change, rare long-lived organisms gaining far more from a year favourable to reproduction than they lose in unfavourable years. Gillett (1962) also points out that a forest may contain many species of trees because insect pests do not allow any one of them to become too common.

Immigration rates and accessible routes for spreading are the last major part of this thesis to explain the species-richness of tropical forests. Their extent and distribution has varied considerably in the past, and it is necessary to consider the likely position of forest refuges, particularly those in the last 2 million years during the Quaternary Period. Haffer (1982) postulates that only small refuges remained in the intertropical zone during the peaks of the glacial periods. These fragmented portions of the ancient populations acted as centres of enhanced speciation and were the source of dispersing spores and seeds.

Thus the flat Amazon and Zaire Basins, lacking any great physical barrier, were easily recolonised by old species or invaded by new forms. Some authors assume that the drier regions were better centres of speciation (Rejmánek 1976). In the case of islands, their distance from the continents or larger islands played an important role, as has been shown by Diamond (1973) for birds. Thus the different contributory factors overlie each other in an intricate web with their relative importance still being hard to appreciate. Much remains to be done to understand the mechanisms that cause and permit the species-richness in tropical forests.

5 Tree growth physiology

In the previous chapter, the structure and ecology of tropical forest communities have been considered as somewhat static structures. Of course these ecosystems are continually changing through growth and reproduction, death and replenishment, and these more dynamic aspects will be covered in the next chapter. Here the physiology of growth and development, especially of the trees, the major component of those communities, will be discussed against the background of the climatic and other factors described in Chapter 3. The emphasis is firmly on whole-plant physiology, subjects such as metabolism, nutrition, translocation and water relations being mentioned only where they have special relevance to the central questions concerning the responses of particular organs.

How does temperature, for instance, influence the production or growth of leaves? What are the factors which control and integrate the allocation of carbon resources to the formation of new organs, against growth or storage in existing parts of the tree? (See Figs 5.1 and 5.16.) Such topics are likely to be of the greatest interest both to biologists and to managers seeking to understand and use tropical forest land, for in general terms they appear to be more specific to the tropics than the underlying processes such as respiration, uptake of mineral ions or protein synthesis.

Since the term growth implies the self-multiplication of living material, it includes in its widest sense all the increases and changes that occur in the life of an individual. However, a convenient distinction is usually made between quantitative increases in size or weight – growth in its narrower sense – and development, which covers qualitative changes such as the initiation of new organs, the starting and stopping of growth, and other relatively abrupt alterations to the existing physiological state. For example, measurements of the rate of expansion of a leaf blade or elongation of its petiole come into the first category; while assessments or observations on the emergence of new leaves, the duration of their period of growth and their ultimate loss from the tree involve the second.

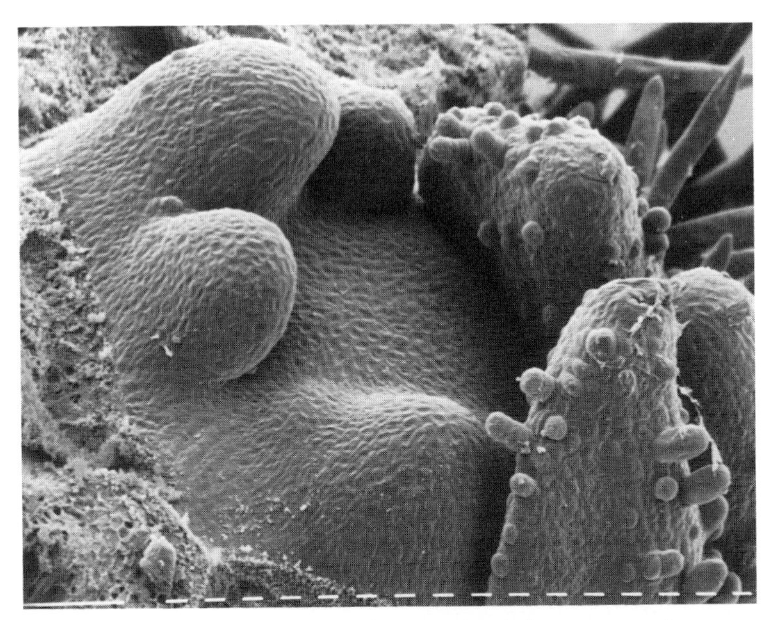

Fig. 5.1 Scanning electron micrograph of shoot apex of cocoa. The very hairy outer leaves and stipules have been removed to reveal a young leaf primordium with its two developing stipules (top left). The next oldest 3-part primordium on the right is beginning to form hairs, while at the bottom of the picture the next youngest primordium is seen, with one of its stipules just visible. (Each dash on scale = 10 μm.)

Thus the size and shape of fully expanded leaves are a function of developmental as well as growth processes. On a given tree, they will vary because the same genetic potential has been expressed in differing environments, and also because of within-plant competition and internal control systems. When comparing with a neighbouring tree, inter-plant competition, age and inherent differences may be involved as well. A logical simplification is therefore to start an experimental investigation by growing potted plants under controlled environments, keeping the conditions as nearly uniform as possible, except for the factor(s) being studied. Since genetic variability between seedlings is generally large, it is often an advantage to work with a known progeny, or better still to develop clones of well-rooted cuttings originating from seedlings (see sections 5.7 and 7.4). Such material can also be invaluable for research on within-tree variation, by allowing the 'temporary' influence of previous position to be separated from the more 'permanent' effects of physiological age or maturity (see section 5.8).

Experiments of this kind cannot usually provide direct information on what factors are controlling growth and development in the forest setting, but this is not their primary function. Essentially, they generate knowledge about the responsiveness of a species to individual environmental and other factors. After a time, however, a sound scientific basis will be built up, from which it may be possible to predict what are the most likely ecological factors operating in the forest, and how they may interact with each other and with internal control systems. Further experiments may then be indicated, or the most appropriate parameters to be used in a theoretical model. In this context, it is important to distinguish between 'ultimate' factors, the various components of the local ecological conditions to which an indigenous species is normally adapted through evolution; and 'proximate' factors, some of which act as 'triggers' or signals that set in train a particular developmental change.

As mentioned in Chapter 1, it is often imagined that all the trees in the tropical forest are producing new leaves and stems continuously throughout the year. The fact that their shoot growth is typically periodic, with longer or shorter intervals of activity and rest alternating with each other has been stressed since the classical studies of Schimper (1898) and Coster (1923). Nevertheless, reference to 'continuous growth in the tropics' still persists, particularly in the more journalistic of articles. On the other hand, Kramer and Kozlowski (1960) go too far in stating that 'no matter how favourable the environment, woody plants do not grow continuously but alternate between flushes or periods of activity or dormancy'. The true position is that seedlings of many species may grow without a pause for a few months or years, but the proportion of woody species which continue to do so afterwards appears to be less than 20 per cent in Java (Coster 1923) and West Malaysia (Koriba 1958). Many of these were shrubs or small trees, and it may well be that the only groups in which continuous growth of large specimens is common in the natural habitat are the tree-ferns, conifers and palms.

Once it is accepted that periodicity of shoot growth is typical of the dominant members of the forest community, the important question follows: are their periods of growth and rest appreciably synchronised or seasonal, or do they occur more or less at random? The latter view was in fact taken by the pioneer Danish ecologist Eugen Warming (1895; 1909), who considered that 'there is no periodicity in the life of the forest as a whole'.

However, subsequent work shows that it is quite usual to find seasonal peaks and troughs in flushing, leaf-fall, flowering and fruiting, not just in climates with a pronounced dry season, but in rain forests where the conditions vary only slightly or irregularly.

Indeed, Lieth (1974) considers that there is virtually nowhere on Earth that can be assumed to have no seasonality of climate. However, in order to show whether trees respond to seasons, repeated observations on tagged shoots over several years is needed, because of the presence of so many species, and considerable variation between trees of different ages, individual specimens and even parts of a single tree. Some of the seasonal, non-seasonal and irregular patterns that have been detected are described in the following sections, which deal with the growth and development of the main organs of tropical trees.

5.1 Bud-break

The first sign that shoot growth is about to start after an inactive period is often an elongation of minute leaves in terminal and some lateral positions. This may be preceded by the enlargement of the bud-scales where these are present (see section 4.2). The subsequent, usually rapid phase of leaf expansion and internode elongation is generally referred to as flushing, and in some species is quite a striking feature of the forest (Fig. 5.2) which is frequently described in the literature. However, it is clear that the important physiological changes leading to renewed shoot extension must occur some time before any visible signs of outgrowth can be detected.

In an early study, Simon (1914) concluded that there was probably no month without any flushing in the tropical forests of west Java, where the climate is fairly constant. The same may well be true also of some more seasonal environments, but it probably reflects the variability of the trees in the forest, rather than their uniformity. Many authors have shown that usually there is increased flushing at certain times of year, and less at others (see, e.g., Taylor 1960; Njoku 1963; Medway 1972; Frankie *et al*. 1974). Some species, particularly in rather uniform climates such as that of Singapore, seem to flush regularly but to be 'out of step' with the calendar months. Others show irregular periods of bud-break, often with a great deal of variability both between and within individual specimens (see Fig. 5.12). Nevertheless, a general tendency towards seasonal flushing appears to be a characteristic feature of the majority of tropical forests.

Fig. 5.2 Flushing vegetative bud of *Brownea* sp. in Singapore Botanic Garden. The pale young leaves all emerge together and hang downwards. Above: older foliage.

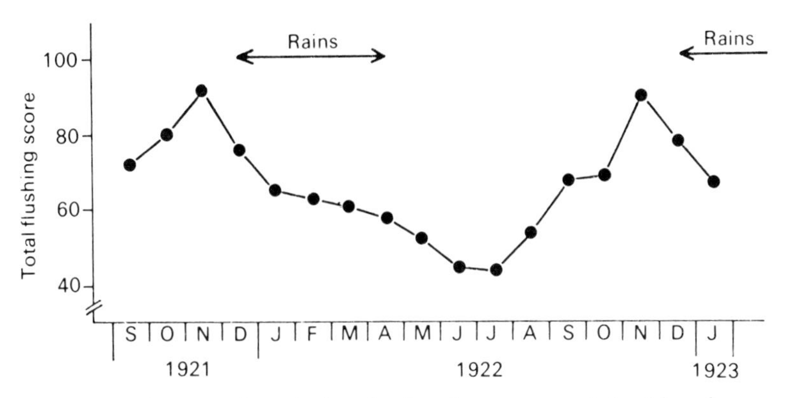

Fig. 5.3 Seasonal variation in the frequency of flushing in a tropical forest at Toeban, W Java. 52 tree species scored on a 0–3 flushing scale. (After Coster 1923.)

An example of a single-peak, annual flushing cycle is shown in Fig. 5.3, which is based on the classical studies of Coster (1928) in Java. Around twice as much flushing was found in November as in June and July, and furthermore bud-break appears to be triggered before and not at the time of the heaviest rains. Alvim (1964) has drawn attention to the frequency with which peak flushing coincides roughly with the equinoxes, in both Northern and Southern Hemispheres and even very close to the Equator. For example, when cocoa plants were grown under partial forest cover, they showed major flushing peaks around the time of both equinoctial periods (Fig. 5.4). Other trees showing similar behaviour include *Terminalia catappa, Peltophorum pterocarpum* and *Cola acuminata*. Some trees show even more frequent recommencement of shoot growth, including cocoa when cultivated in full sunlight where the conditions are less comparable with the natural habitat of this small tree of the forest undergrowth.

Bud-break can therefore occur at any time of year, but is more likely in certain months. The pre-rains flushing already referred to is frequently though not invariably found, and flushing peaks just after the wettest season also occur (Whitmore 1975). At first sight, this seems surprising, since the rains are so obviously the growing season for many crop plants, grasses and other herbaceous species. However, it is evident that for tropical forest trees as a whole this assumption may not be correct. The 'ultimate' factors which may perhaps have influenced the evolution of such seasonal replacement of the majority of the forest's assimilating surfaces will be further considered in section 5.3, but

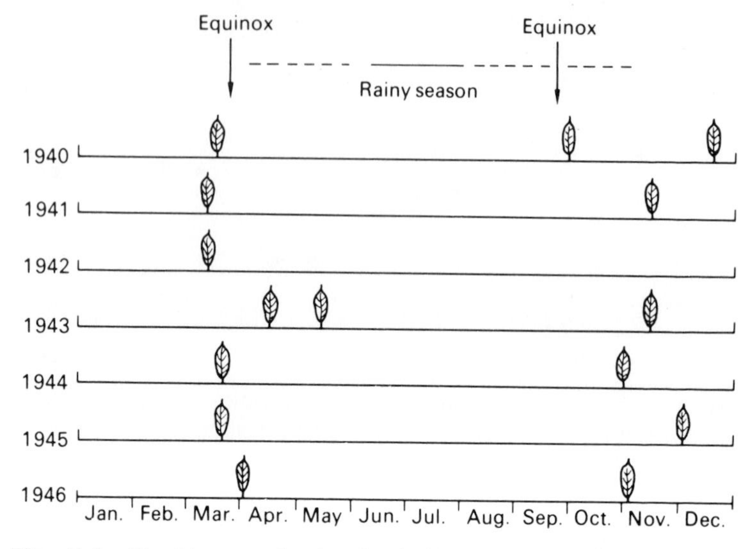

Fig. 5.4 Flushing peaks in shaded cocoa in plantations near Tafo, Ghana (6 °N). Symbols represent times when 50 per cent or more of trees were flushing. Time of bud-break estimated at about 2 weeks earlier. (After Greenwood and Posnette 1949.)

attention will now be focused on the question of what 'proximate' factors may trigger bud-break.

It is often tacitly assumed that renewed flushing is caused by rainfall, but this clearly cannot be correct in the common situation when outgrowth starts in dry weather. Indeed, even for those tree species that flush after rain, there is often no more than circumstantial evidence that water is directly involved. Njoku (1964), for instance, has warned of the dangers of trying to fit crop growth measurements to rainfall data and infer a causal relationship. Correlations alone, no matter how high their levels of statistical significance, only provide the working hypotheses that need experimental or theoretical testing if they are not to mislead.

In some species, evidence indeed exists that increased **water potential** (decreased water stress) may promote flushing. In teak, for example, Coster (1923) found that simply standing cut leafless shoots in pure water was usually enough to bring about bud-break in 7–10 days, whether they were inside a room or out-of-doors. Similar effects were found in preliminary studies by Njoku (1963) with *Terminalia superba, Bosqueia angolensis* and *Millettia thonningii*. The most likely explanation for these results is that the buds were in a state of quiescence or 'post-dormancy' (Wareing 1969), requiring only an increase in water content to flush, rather as a seed may lack just water for germination.

However, the same authors found that cut twigs of *Bombax malabaricum* and *Sterculia tragacantha* could only be forced in water when taken either early or late in the rest period. Thus these species may resemble many woody plants of the temperate zone in showing a period of 'true dormancy' during which the terminal buds will not flush even though conditions are favourable for shoot extension growth. (Lateral buds can be subject to internal control by terminals; see section 5.8.) The early and late portions of the rest period, in which outgrowth of new shoots can occur under favourable conditions, are referred to respectively as 'pre-dormancy' and 'post-dormancy'. Much more research is needed before the role played by water can be properly assessed, but two points may nevertheless be made. Firstly, many observations and experiments with well-watered potted seedlings, cuttings and mature grafts indicate that intermittent growth and rest can still occur, suggesting that other controlling factors are involved. Secondly, rainfall could have effects other than relieving water stress, such as the leaching of water-soluble inhibitors from the buds, or indirectly through the sudden temperature drops associated with tropical rain-storms.

In many ways **temperature** is the most likely environmental factor to be involved in bud-break. It is perceived by all living cells, and affects virtually every chemical reaction, as well as the physical background for life processes. It is also known to have overall developmental effects: in many temperate trees, for example, chilling during the winter at temperatures just above freezing point breaks the true dormancy of their buds, while rising temperatures in the spring control their flushing in the post-dormant condition. It is not yet clear whether there are analogous systems in tropical trees, although in *Acer rubrum*, which has a very wide range from eastern Canada to south Florida, the tropical ecotype has no chilling requirement for bud-burst (Tomlinson 1980). In cocoa (which produces repeated short cycles of shoot growth) mature cuttings flushed much more frequently in growth rooms at 30 ° or 31 °C than at 23 °C (Murray and Sale 1966; 1967). Day temperatures appeared to be rather more important than those at night, but there was no suggestion that diurnal fluctuations as such were involved. Earlier correlations of flushing frequency with field temperature measurements had suggested that bud-break might be promoted when maximum temperatures exceeded about 28 °C, and by diurnal fluctuations greater than 9 °C.

A more unexpected factor requiring consideration in the tropics is **photoperiod**. It has often been assumed that because the changes in day-length occurring near the Equator are relatively small, plants might well be insensitive to them. However, it is equally possible that they might be more sensitive to small changes than species of the temperate zone, and indeed some rice varieties have been claimed to detect differences of 5 minutes (Stubblebine *et al.* 1978). It is also known that some tropical animals are sensitive to photoperiod. Regarding woody plants, Alvim (1964) suggested that cocoa at 10–15° S showed a period without flushing because the days were shorter at that time. However, increasing day-lengths in a plantation to 15–16 h by means of bright lights did not stimulate flushing (Alvim and Grangier 1965). In unpruned tea bushes growing at about 27° N, on the other hand, promotion of bud-break did occur when the short natural day-lengths of November – March were extended to 13 h with weak supplementary illumination (Barua 1969).

Experiments in controlled environments with various tropical forest tree species have shown clear sensitivity to photoperiod (Longman 1969). Leafy seedlings of *Hildegardia barteri* remained dormant and did not flush under short-days (9 h 10 min). However, bud-break did occur when most of the leaves had become senescent and/or had been abscised, or where the plants had been leafless at the beginning of the experiment. Under long-days (17 h 10 min), flushing took place whether leaves were present or not. Bud-break in mature cuttings of *Cedrela odorata* was also prevented by short-days unless the mature leaves had been shed. Interestingly, *Bauhinia acuminata* plants that showed earlier leaf senescence under short-days also resumed shoot elongation more quickly than the still leafy seedlings growing under long-days (Singh and Nanda 1981).

It would appear that the buds in these experiments were predormant; prevented from flushing through inhibition by the mature leaves (Wareing 1969). They could be released either by transfer to a favourable photoperiod or by loss of the leaves. In this connection, it is interesting to note that out-of-season flushing can be stimulated when *Brachystegia laurentii* is defoliated by caterpillars (Germain and Évrard 1956) and that experimental defoliation induced bud-break in *Couroupita guianensis* grown in a glasshouse at Heidelberg, Germany (Klebs 1926).

Besides these environmental factors and the inhibitory influence of mature leaves, it is likely that in some species endogenous rhythms are involved (Vogel 1975; Borchert, 1978), though it is not necessary to invoke general hypotheses of long-term, yearly rhythmic cycles to account for the regular annual occurrence of bud-break. The internal control systems of cocoa have been particularly studied, and it has been found that periods of rapid root growth alternate with shoot flushes (Sleigh *et al.* 1984). When portions of the root system were removed, bud-break was delayed, suggesting that the close relationship between root and shoot activity might involve growth substances. Moreover, application of abscisic acid (ABA) to the leaves of the recent or previous flush also delayed bud-break. When ^{14}C-ABA was applied to the leaves, it accumulated at the shoot apex, particularly during the dormant period (Hardwick *et al.* 1982). Conversely, applying the gibberellin GA_3 or the cytokinin zeatin promoted flushing, the radioisotopes of these growth promoters accumulating at the apex most markedly during bud-burst and leaf expansion.

Further research is needed on internal factors, which might for instance control bud-break via the leaf-exchanging habit (section 5.5), or prevent flushing in species where the vegetative buds do not grow out while a branch is flowering, such as *Ceiba pentandra* and *Hildegardia barteri*. Carbohydrate levels may be important as well as growth substances, for Fink (1982) has shown that the starch levels in stemwood of ten Venezuelan tropical tree species declined during flushing and increased again afterwards.

5.2 Rate of stem elongation

Two processes are involved in the production of new stems: the formation of additional nodes and the elongation of the internodes between them. Sometimes the new nodes (together with foliage leaf and bud-scale primordia) remain as a resting bud after their initiation by the shoot apical meristem, and internode elongation occurs later on. The stem units produced in a flush shoot may all have been pre-formed, or the apex may continue to form additional nodes after bud-break, so that 'fixed growth' is followed without a break by 'free-growth'. When the shoot is growing continuously for an extended period, new nodes are being formed and internodes extending without a break. This section is concerned with the many factors affecting the rate at which these processes occur, including the production of new leaves. Those influencing leaf expansion, which usually takes place at the same time, are considered in section 5.4.

The rate of stem elongation in some tropical woody plants can be extraordinarily high. Certain species of bamboos may even elongate pre-formed tissue at almost 1 m per day, and vines and lianes can also grow very fast. On a longer time-scale, Fig. 5.5

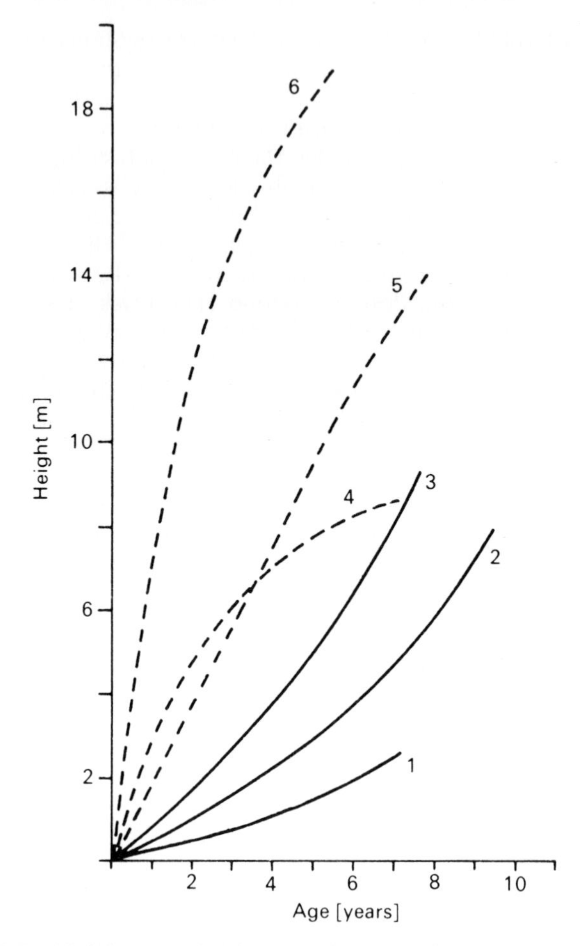

Fig. 5.5 Height growth of young trees under natural conditions in Zaire. Full line = **canopy** species: (1) *Scorodophloeus zenkeri*; (2) *Oxystigma oxyphyllum*; (3) *Gilbertiodendron dewevrei*. Dotted line = **secondary** species in clearings: (4) *Calancoba welwitschii; L* (5) *Terminalia superba*; (6) *Musanga cecropioides*. (After Lebrun and Gilbert 1954.)

shows early growth rates of young trees under natural conditions in Zaire, the three climax species increasing in height at an average rate of 0.3–1.2 m per year, and the three pioneers at 1.2–3.7 m per year. It is also noticeable that the former group increased their height increment as they got older, whereas rates steadily declined in the latter. In their study of over 1 500 trees during early succession of cleared forest in Ghana, Swaine and Hall (1983) found that after 5 years the majority of the late seral stage trees had grown at less than 0.5 m per year. Conversely, there were 8 species of pioneers that had exceeded an average rate of 2 m per year. Rates of height growth in the first year can be as much as 3–4 m for such heliophilous species, when thriving as seedlings or coppice shoots in large clearings or plantations. Exceptionally, 8–9 m in the first year has been recorded in *Albizia falcataria* (Nicholson 1965), *Sesbania grandiflora* (NAS 1979), *Eucalyptus deglupta* and *Trema micrantha*, the latter exceeding 30 m in 8 years (Mabberley 1983).

Although height growth in tropical trees can clearly be very rapid indeed, it is easy to exaggerate the general growth rates in the forest by concentrating attention on the performance of the leading shoots of a few species in particularly favourable conditions. Most of the other shoots on a tree grow more slowly, and by the time the middle and upper tree levels are reached, stem extension, even of the most vigorous shoots, may well be reckoned only in centimetres per year. At the lower tree level, woody plants often survive with minimal increase in height for many years, while growth in small gaps is usually much less than in the open.

The rate of stem extension can be affected by all the environmental factors discussed in Chapter 3, as well as by others such as the force of gravity (Damptey 1964) and the presence of atmospheric pollutants. Not surprisingly, **water availability** is important, rates often being affected by quite small deficits. Currently expanding internodes can be inhibited, and reduced numbers of nodes may be formed during rest periods. Cell expansion in plants clearly depends directly on turgor, while photosynthesis, metabolism and growth often slow down because water deficits stimulate an abscisic acid-induced closing of stomata that may persist for some days, even if the stress has been relieved.

Daily fluctuations in growth rate of stems, as well as longer term effects, can result from alterations in plant water potential. Sixty years ago, Coster (1927) used an automatic recording technique to follow in detail the elongation of a young tropical bamboo shoot. During the night it grew at 13 mm h^{-1}, but there was a very pronounced check to extension growth during a hot day. The following day was also hot and sunny, and by 11.30 h the rate had fallen to 5 mm h^{-1}. At this point, he removed the four tall mature canes growing from the same clump, whereupon before noon the growth rate increased to 16 mm h^{-1}. Since the change was so large and so rapid, the conclusion is almost

Fig. 5.6 Effect of day-length and night temperature on the shoot growth of (A) *Terminalia ivorensis*; (B) *Triplochiton scleroxylon* seedlings. From the left, row (1) 11 h (20 °C nights); row (2) 14½ h (20 °C); row (3) 11 h (30 °C); row (4) 14½ h (30 °C). Day conditions similar for all treatments. Note the interactions: faster shoot extension in row (4) (A); reduced leaf expansion in row (1) (B).

inescapable that elongation had been retarded by temporary water stress, and that it increased because the main transpiring surfaces had been removed. In contrast, Coster found that the vine *Aristolochia gigas* grew at a rather steady rate throughout a night and a hot day, while in the liane *Congea villosa* day-time growth exceeded that made in the night. In view of such diversity of response, it would be unwise to make any general assumptions that shoot growth is typically greater at night, as is sometimes done.

Diurnal fluctuations of **air temperature** in the tropical forest can sometimes be considerable, particularly in the upper canopy (section 3.2). Seasonal changes are often less pronounced or minimal, and comparison with temperate zone conditions has led some to conclude that temperature might not be a particularly important factor. However, experiments with well-watered potted plants in controlled environments indicate quite clearly that many tropical tree species are particularly sensitive to small temperature differences. Thus *Ceiba pentandra* seedlings kept at 36 °C grew 23 times faster than others from the same batch placed at 15 °C (Kwakwa 1964). In several experiments, increasing only the night temperature by 5–10 °C promoted shoot elongation rates substantially (Longman 1978; Fig. 5.6 and Table 5.1). Cocoa appears to be a particularly sensitive species, for in one trial raising only the day temperature by 3.5 °C stimulated shoot growth of 250 per cent (Murray and Sale 1966).

The full range of temperatures in which shoot growth occurs has not been thoroughly explored. Minimum temperatures may often lie in the range 12–15 °C (Larcher 1980), below which many tropical plants in any case show chilling injury (see section 3.2). *Guarea trichilioides* and *Avicennia marina* appear to have a high minimum temperature of 21 °C for shoot growth (Altman and Dittmer 1973). Maximum temperatures for shoot elongation (and for survival) often seem to lie between 40 °C and 50 °C. As a general rule, the optimum temperature for shoot growth lies nearer the maximum than the minimum end of the range. A value of around 36 °C was suggested for young seedlings of *Ceiba pentandra* and *Leucaena leucocephala* (Kwakwa 1964), which is perhaps rarely likely to be experienced in nature.

Another feature of the shoot elongation of *C. pentandra* is that it is more rapid with constant rather than fluctuating temperatures (Fig. 5.7). Moreover, if the pairs of shaded and solid histograms are compared, it is clear that interchanging the day and night conditions does not make a great deal of difference. However, in the small woody coastal plant *Cassia mimosoides*, the warm day/cool night regimes were generally more favourable,

Table 5.1 Effect of night temperature on shoot elongation of seedlings of tropical forest trees

Species	Night temperature (°C)			
	Lower	Higher	Rise in temperature	% Increase in shoot growth
Cedrela odorata	20	30	+10	+ 94***
Triplochiton scleroxylon	20	30	+10	+ 65***
Ceiba pentandra	26	31	+ 5	+ 74***
Ceiba pentandra	31	36	+ 5	+ 6 n.s.
Bombax buonopozense	26	31	+ 5	+ 72***
Bombax buonopozense	31	36	+ 5	+ 21*
Gmelina arborea	26	31	+ 5	+140***
Gmelina arborea	31	36	+ 5	− 14 n.s.

The seedlings were grown in soil in pots, and were from 1 to 15 months old at the beginning of the experiments. There were between 10 and 14 plants in each treatment, and they were kept out-of-doors or in a lightly-shaded greenhouse during the day, and in the appropriate growth-room at night. The night temperatures were in operation for between 9½ and 13 h, and treatment lasted for 5–21 weeks. Significance of increases in Tables 5.1 and 5.2:
*** – 0.1% level
** – 1% level
* – 5% level
n.s. – not significant

and alternating temperatures often led to more growth than constant conditions. The optimum regime appeared to be 29 °C day/22 °C night, close to conditions that might often be experienced in nature. Similarly, young coffee plants made most shoot growth with the combination 30 °C day/23 °C night (Went 1957).

Although cooler nights may well conserve assimilates, it should not be assumed that most species will necessarily be similar to the two just described. Even in *C. mimosoides*, when the day temperature is kept at 15 °C or 22 °C, shoot growth is promoted when nights are 7–14 °C warmer than the days (Fig. 5.7). *Gmelina arborea* apparently has a night temperature optimum for leaf production near to 36 °C, while it is probably around 31 °C for internode elongation of *Ceiba pentandra* (Longman 1978). Some species actually have a higher night temperature optimum than that for the day, such as the tropical herb *Saintpaulia ionantha* (Went 1957), and some provenances of the subtropical conifer *Pinus ponderosa* (Callaham 1962). One way in which warm or hot nights might perhaps stimulate shoot growth would be if they promoted leaf production and/or expansion, and thereby increased the plant's subsequent capacity for photosynthesis.

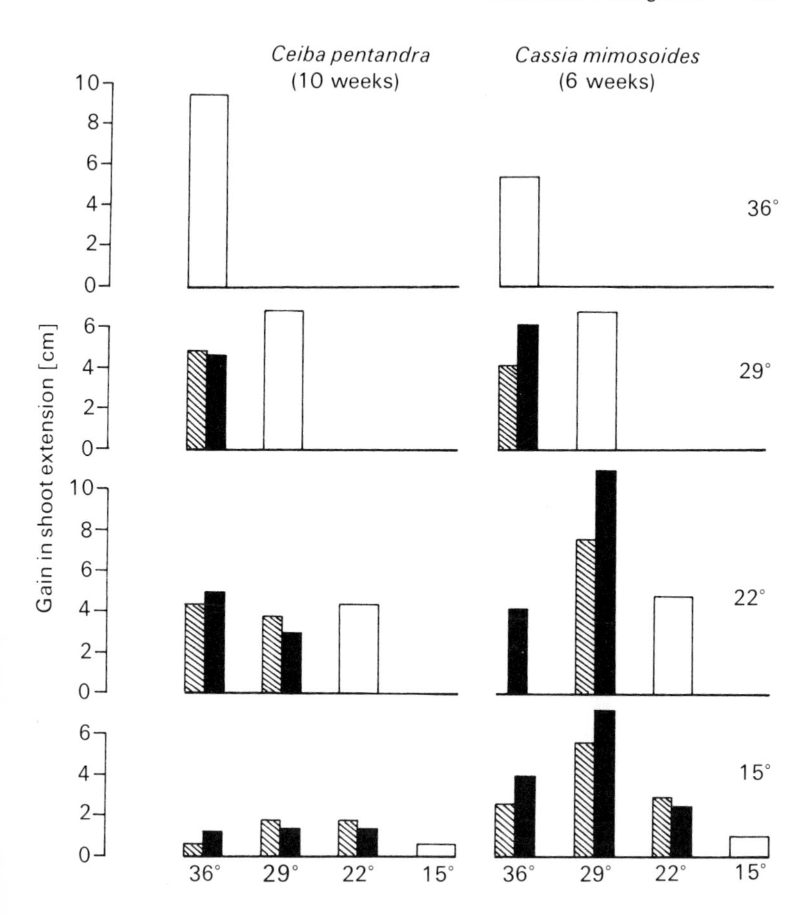

Fig. 5.7 Effect of constant and fluctuating air temperatures on the rate of shoot elongation in *Ceiba pentandra* and *Cassia mimosoides* seedlings. **Open** columns = constant temperatures; **solid** columns = fluctuating temperatures, with **cooler** night; **shaded** columns = fluctuating temperatures, with **warmer** night. Both day and night regimes lasted 12 h; day-length = 13.2 h. Almost all *Cassia* plants in 22 °C day/36 °C night died before week 6. (After Kwakwa 1964.)

The rate of elongation by shoots may also be affected by the **photoperiod** under which the tree is grown (Stubblebine *et al.* 1978). Not only are at least 14 tree species sensitive to this factor, but the effects can be quite large (Table 5.2). For instance, shoot growth in *Terminalia superba* seedlings was trebled by increasing

Table 5.2 Effect of day-length on shoot elongation of seedlings of tropical forest trees

Species	Photoperiod (h)			
	Shorter	Longer	Increase in day-length	% increase in shoot growth
Cedrela odorata	11.0	14.5	+3½	+ 51**
Theobroma cacao	12.5	14.5	+2	+ 31**
Terminalia superba	9.2	13.2	+4	+195***
Terminalia superba	13.2	17.2	+4	+ 59**
Chlorophora excelsa	9.2	13.2	+4	+ 22 n.s.
Chlorophora excelsa	13.2	17.2	+4	+355***
Ceiba pentandra	9.2	13.2	+4	+130***
Ceiba pentandra	13.2	17.2	+4	− 1 n.s.
Bombax buonopozense	9.2	13.2	+4	+ 82***
Bombax buonopozense	13.2	17.2	+4	− 7 n.s.
Hildegardia barteri	9.2	17.2	+8	+200***

All plants received approximately the same total light energy, the extension of the day-length being given with low intensity illumination (20–200 lux). The maximum day-length 5 degrees from the Equator is approximately 13.2 h, allowing for dawn and dusk. Between 11 and 20 plants were used in each treatment; other details as Table 5.1.

the day-length from just over 9 h to the value found in June near the coast of West Africa. This species is so sensitive that statistically significant photoperiodic effects could be detected after as little as three days' treatment. The effect involved both internode elongation and the number of nodes and leaves produced. The latter shoot growth parameter, and also shoot and total dry weights, were especially stimulated in *Chlorophora excelsa* when day-lengths were lengthened to about 17 h (Longman 1978).

All the treatments in these experiments received similar photon flux densities of shaded natural sunlight in glasshouses and fluorescent light in growth rooms, the day-lengths being extended as necessary by supplementary incandescent illumination too dim to influence photosynthesis appreciably. Thus it is clearly established that many tropical trees are sensitive to photoperiod, influencing either the rate of leaf production or internode extension, or both. Other species in which long-days promote shoot growth include *Pinus caribaea* var. *hondurensis, Rauvolfia vomitoria, Hymenaea courbaril, Bauhinia acuminata*, coffee and cocoa; while extension of *Gmelina arborea* shoots appears to be unaffected by changes in day-length.

Synergistic effects between day-length and night temperatures have been found in *Terminalia ivorensis* (Fig. 5.6A), and in *Ceiba*

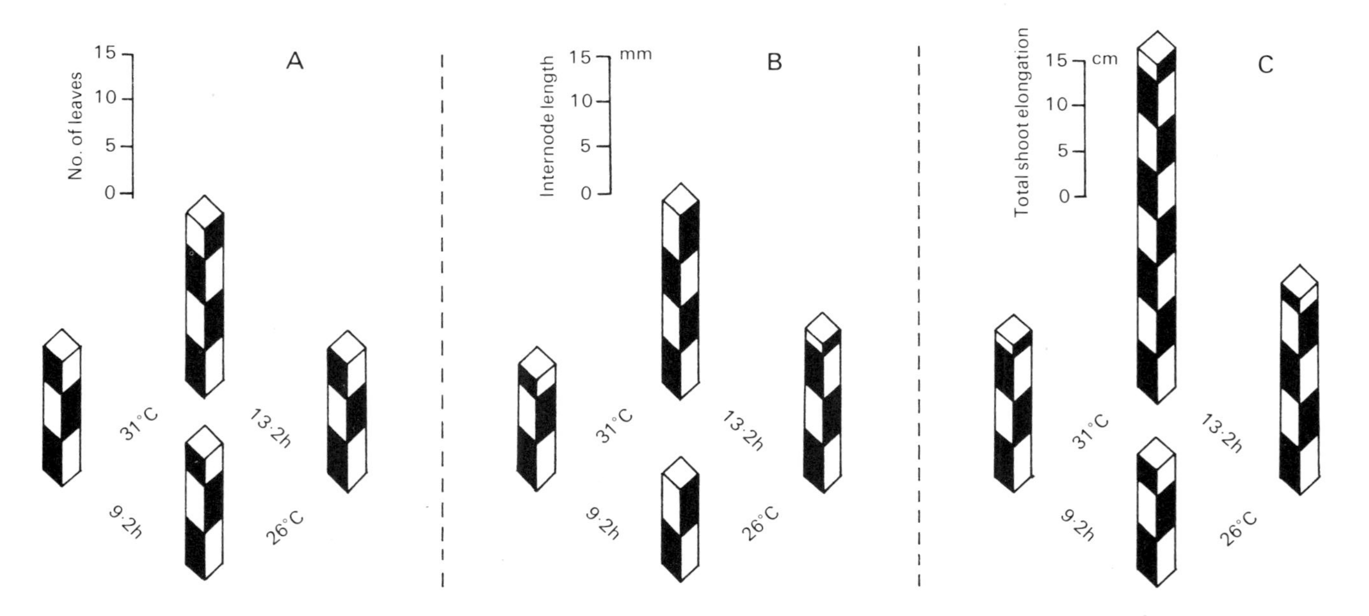

Fig. 5.8 Interactions between day-length and night temperature affecting the rate of shoot elongation in *Ceiba pentandra* seedlings over a 21 week period. Note that here the gain in height (C) involved both more leaves (A) and longer internodes (B). Day temperature: 31 ± 3 °C.

pentandra where 50 per cent more leaves were formed when both promotive factors were applied, but neither had a significant effect when given alone (Fig. 5.8). More complex interactions appeared to be involved in the control of internode length and total shoot elongation, which were both strongly influenced by day-length and night temperature. Thus the question as to whether trees under natural conditions will be influenced by photoperiod is difficult to answer. However, it is clear that shoot growth of a number of species can be substantially modified, increases of up to 75 per cent per hour having been demonstrated (Table 5.2). Some evidence exists of sensitivity to alterations in the range experienced in the humid tropics, where natural day-lengths vary by just over 30 minutes at a latitude of 5 °C, and by more than an hour at 10 °C. Besides possible photoperiodic effects, the shoot growth of forest trees may be affected by the light quality received, especially by the red/far-red ratio (Kwesiga 1985). The photon flux density is clearly also an important consideration, both indirectly through carbon assimilation and perhaps also directly. Effects of shading on growth are often profound, as has been shown for instance with seedlings of several species of Dipterocarpaceae (Sasaki and Mori 1981). It should be remembered that differences may not always be due to light itself, since treatments may also produce modifications of light quality, leaf temperatures and water potential.

Clearly, many other factors will affect the rate of shoot growth. For instance, there was a very high correlation ($r = 0.99$) between the level of phosphorus fertilisers added and the height growth of *Terminalia ivorensis* seedlings (Aluko and Aduayi 1983). Besides external conditions, it is important to bear in mind that other processes which are proceeding simultaneously in the tree, such as root growth, may affect the shoots, as pointed out more than 25 years ago by Kursanov (1960) regarding the aerial roots of an epiphytic *Ficus* sp. Flowering or fruiting may divert carbohydrates away from shoot growth, or perhaps modify it through the mediation of growth substances. It is also possible that sizeable differences may arise through the presence or absence of specific mycorrhizal fungi associated with the root system (Ivory and Munga 1983; see also section 6.3). Genetic differences between individuals in a population or species can be surprisingly great, as is demonstrated by the very substantial clonal variation in height growth that is typically found between clones of tropical forest trees (Longman 1982; Ladipo *et al.* 1983), which in *Triplochiton scleroxylon* may be related in part to different net assimilation rates (Ladipo *et al.* 1984). Higher photosynthetic rates per unit leaf area have also been reported for *Ceiba pentandra* than for *Musanga cecropiodes, Terminalia ivorensis* or *Chlorophora excelsa* (Okali 1971).

5.3 Onset of bud dormancy

It is often possible to detect that a flush or a longer period of extension growth is coming to an end by observing that there are few or no expanding leaf primordia at the shoot tips (Fig. 5.9). Once the terminal bud has entered some form of dormancy, stem elongation is confined to the completion of the extension of the last few internodes. The shoot apex itself is not necessarily inactive, and often continues to initiate new leaves and 'stem units' during the dormant period, at least for some time. Besides foliage leaves, the apex may produce bud scale primordia; and there are some species in which dormancy is signalled by the abscission of the whole of the terminal shoot tip, leaving lateral buds to continue axial growth. Such shoot tip abortion is a common feature of the shoot growth of *Cedrela odorata, Xanthophyllum curtisii* and *Citrus* (Addicott 1982). Thus the height of a tree may depend on three additional considerations, besides the factors influencing the rates of node production and internode elongation discussed in the previous section. These are the **duration of the period of shoot elongation**, the **number of preformed nodes within resting buds** and the **duration of the period of bud dormancy**.

Fewer ecological studies have covered the time of cessation of shoot elongation than the more obvious phenomenon of flushing. What information there is suggests that in plantation crops that grow by repeated flushes, the duration of elongation may be 6–7 weeks, for example in cocoa, and as little as 1½–2 weeks in rubber and mango. In the forest, it seems likely that periods of shoot extension of 1–2 months may become fairly general after the seedling and sapling stages (Njoku 1963; Alvim, 1964). The onset of bud dormancy will therefore occur at different times of year in trees which flush intermittently or irregularly, but in more regularly flushing species it will tend to be seasonal.

In Ghanaian forests, for example, the commonest month for the ending of shoot elongation appears to be April (Taylor 1960), which is in the transitional period between the main dry and rainy seasons. Even in a species such as cocoa, which may flush several times a year (Fig. 5.4) it was observed several decades ago that

> *most of the growth of mature cocoa occurs in the drier months . . . while during the main wet season when conditions of water supply and humidity are most stable, little or no growth occurs* (Greenwood and Posnette 1949).

Fig. 5.9 Effect of day-length on terminal bud dormancy in *Cedrela odorata* seedlings: (A) actively growing shoot apex in long-days; (B) dormant apex under short-days.

In a tropical dry forest in Costa Rica, production of new leaves and shoots declined as the rainy season advanced, and in a wet evergreen forest it decreased as the drier part of the year passed (Frankie *et al*. 1974).

Although there are exceptions, it is remarkable that many tree species should be ceasing shoot elongation at a time when the 'growing season' is just starting. There is a parallel here, however, with the situation in the north temperate zone, where many species cease producing new leaves in May, June or July, when the environment for growth seems ideal. As was pointed out a century ago (Sachs 1887), trees often re-start shoot growth under less favourable temperature and other conditions than those obtaining when they stop elongation. However, perennial plants might well gain an important selective advantage by producing new foliage in the latter part of the dry season. Unlike annual grasses and many short-lived crops, their recently expanded leaves, at the peak of their photosynthetic capability, will then be present at the time of the greatest daily totals of photon flux density. Light levels are usually reduced by greater cloudiness in the rainy season (see Fig. 3.7), and sometimes by dust or haze at the height of the dryest season, giving sunlight maxima around the equinoctial periods. An additional 'ultimate' factor that could be operating is that a leaf expanding in the latter part of the dry season may be subject to less overall water stress than one formed at any other time.

If the dormant period in many mature trees is as long as 10–11 months, this will curtail the yearly height increment even when the rate of elongation is rapid. In seedlings too, growth of *Triplochiton scleroxylon* and *Hildegardia barteri* fell behind that of *Musanga cecropioides* mainly because they became dormant for a while, when the *Musanga* grew continuously (Coombe and Hadfield 1962). This may also be one reason for the decline in height growth often found with increasing age: for example 3-year-old *Bombax buonopozense* stop height growth for only 3–4 months, whereas older trees are dormant for 9–10 months (Njoku 1963).

The physiological signals which induce a change from active shoot growth to terminal bud dormancy appear to include both external and internal factors. For instance, it has been demonstrated using controlled environments that shoot extension can be halted in a number of tropical tree species by reducing the **photoperiod**. Seedlings of *H. barteri* stopped extension growth quickly under 11½ h days, and more slowly in 12 h days. When given 12½ day-lengths, however, they grew continuously throughout the 26 week experimental period (Njoku 1964). Very rapid onset of dormancy also occurred in leafy *Cedrela odorata* plants exposed to short-days (Fig. 5.9), but the process was slower in *Terminalia superba* seedlings, about 50 per cent of the treated trees stopping temporarily, while the other half continued to grow slowly (Longman 1969; 1978). A proportion of *Ceiba pentandra* seedlings grown under short photoperiods (Fig. 5.8) also showed temporary dormancy. The length of the shoot extension period was decreased by about 6 weeks in *Bauhinia acuminata* when day-lengths were shortened (Singh and Nanda 1981), but a reduced daily light energy supply or temperature differences may have been involved as well as photoperiod in this experiment using a thick tarpaulin screen to apply the short-days.

Cooler night **temperatures** tended to induce bud dormancy in *Gmelina arborea*, some plants stopping height growth with nights of 26 °C, but none doing so at 31 ° or 36 °C. Cool nights can interact with short-days, as in *Bombax buonopozense* (Longman 1969). Under natural conditions in the forest, a reasonable hypothesis might be that warmer nights and longer days might promote faster and more prolonged shoot extension, while cooler nights and shorter days could lead to reduced shoot growth and to cessation of height growth. However, it is very probable that other external factors will hasten the onset of terminal bud dormancy, for example reduced light intensity, a low red/far-red ratio, water stress and shortage of mineral nutrients.

Internal conditions within the tree also appear to be important in controlling the onset of dormancy. Species which show repeated flushes of growth, such as cocoa, tea, mango and rubber, are good examples of this, although it should not be assumed that they will be unaffected by their external environment. One possibility is that a rapidly growing shoot might stop simply because new leaves were not being initiated quickly enough by the apex, which may happen when tea is grown in the tropics (Bond 1945). Alternatively, each shoot flush may be confined to those leaves and stem units pre-formed in the bud. In this case, dormancy is as it were predetermined, perhaps with an 'articulate' shoot morphology reflecting the cyclic production of bud scales and foliage leaves (Hallé *et al.* 1978).

A series of detailed studies on the photosynthetic capacities of different leaves of cocoa, and the translocation of assimilates within the plant, has shown that when first-year plants are flushing rapidly and expanding many new leaves, they are using much more carbohydrate than they are producing (Hardwick *et al.* 1985). Thus it is possible that lack of assimilates might lead to cessation of shoot elongation; for example, successive removal of expanding leaves can prolong shoot elongation periods, for instance in rubber (Hallé and Martin, 1968) and in *Couroupita guianensis* (Klebs 1926). In temperate species of the tropically originating genus *Quercus*, it is even possible to turn an intermittently flushing tree into one which grows continuously (Longman and Coutts 1974). Some instances are also known in which individual specimens of intermittent tropical species exhibit continuous growth for long periods, as in vigorous leading shoots of unplucked tea. Examples such as 'lamp-brush' rubber and 'fox-tail' pine may be due to genetic differences, perhaps becoming expressed when the plant is grown out of its natural environment.

It is likely that there are often negative feed-back control systems that change or maintain the phases of shoot elongation or terminal bud dormancy. The interactions between roots, leaves and buds may well involve changes in the endogenous levels and distribution of both growth promoters and inhibitors, and changes in levels of these have been detected at different times in the flushing and dormant cycles of cocoa (Hardwick *et al*. 1982). In this species, internal water stress was apparently not involved (Hardwick and Collin 1985), but it has been implicated in studies with other tropical trees (see, e.g., Reich and Borchert 1982).

5.4 Growth of leaves

Young, growing leaves quite often attract attention in tropical forests because their colour or appearance is different from that of the mature foliage. They are often a lighter shade of green, because the formation of chloroplasts and the synthesis of chlorophyll a and b has only just started. Sometimes they are very pale – for example in the half-expanded flush shoot of cocoa grown under shade – and they can even appear white, yellowish or blue in some species (Richards 1952; Hallé *et al*. 1978). The commonest coloration is red, generally caused by anthocyanins in cell vacuoles, and varying from a barely distinguishable pale shade to a bright colour which masks any green present, as seen for example in *Cynometra ananta* and *Carapa procera*. In some species, such as *Amherstia nobilis*, the new flush of pale-coloured young leaves all hang downwards for some time (see also Fig. 5.2), which may be an adaptation that reduces water stress until the leaves are capable of active photosynthesis.

A feature of tropical trees which is fairly common is the capacity for active leaf movements. The orientation of leaves and leaflets can be altered, not just by growth curvature of the petioles, but at swollen **pulvini** or 'leaf-joints' at the base of the organ in question (Fig. 5.10). The turgor changes which are

Fig. 5.10 Leaf-joints (pulvini) at the leaf bases in *Khaya ivorensis*.

responsible for the movement can be relatively rapid, so that the plant reacts quite promptly to a change in environment. The best-known example is the 'sensitive plant', *Mimosa pudica*, which is a cosmopolitan weed of open spaces in the tropics. Here the reaction time to the stimulus of a leaflet being touched is of the order of a second or less, and the leaflets also close (more slowly) in darkness. Light at wavelengths of 403 nm (violet) or 726 nm (far-red), but not at 585 nm (yellow) or 656 nm (red), has been found to cause opening of leaflets, but this was only effective when given between 0600 h and 1600 h, indicating the existence of an endogenous diurnal rhythm. Applied indole acetic acid also led to opening, even in darkness, and autoradiography showed that it could spread from the rachis throughout the pinna in 4 minutes (Watanabe and Sibaoka 1983). In *Albizia*, a red/far-red reversible system has been shown to control opening and closing of the leaflets, the turgor changes in the pulvinus apparently being triggered by a massive influx or efflux, respectively, of potassium ions.

In many leguminous trees, the leaflets close at night, and sometimes also in response to other stimuli – for instance those of *Brachystegia spicaeformis* have been shown to close during the day when leaf surface temperatures surpass 32–34 °C (Ernst

1971), while dark closure was promoted and the duration of the closed phase increased in *Cassia fasciculata* when temperatures were reduced from 35 ° to 15 °C (Gaillochet and Gavaudan 1974). Movements are not confined to the Leguminosae, being noticeable at night in *Triplochiton scleroxylon*, for example, while an unusual example is the formation of 'night-buds' in *Espeletia schuttzii* at 3600 m in the Venezuelan Andes. The rosette of leaves close around the apical buds at night, which has been shown by measurement with thermocouples to protect the young leaves from freezing temperatures, and from rapid heating up just after sunrise, either of which can be lethal to them, but not to the older leaves (Smith 1974).

The physiological significance of changing leaf orientation probably lies in the achievement of a successful balance between maximal photosynthesis and minimal water stress. In addition, as noted in section 4.2, it has important ecological effects on other leaves, by altering the proportion and the quality of light that reaches them at different times. Similar considerations apply to the regular occurrence of upper canopy leaves that are more or less permanently oriented at a steeply inclined angle, rather than horizontally (Medina 1983). The leaf mosaic can also change rather abruptly under a previously leafless tree crown that suddenly produces a new flush of leaves, and there are continual, slighter modifications resulting from the expansion and loss of individual leaves, or grazing by herbivores.

Seasonal changes which affect the forest as a whole may also occur. Peaks of new leaf growth generally occur shortly after those for bud-burst (see section 5.1), because the expansion of leaves typically occurs at the same time as internode elongation. Sometimes the two processes appear to be closely linked; thus, for example, in tea a single deciduous bud scale amongst expanding foliage leaves will have a correspondingly short internode proximal to its insertion. Often the first foliage leaves on a new shoot are smaller (Fig. 5.11), and may be similarly associated with shorter stem units. On the other hand, the first internode of a sylleptic branch (Tomlinson 1978; Fisher 1978) is typically much longer than the subsequent ones, but the leaves are usually of a similar size. In *Terminalia ivorensis*, for example, experimentally reducing the area of young expanding leaves did not significantly affect the internode lengths (Damptey 1964).

Growth rates of leaves are often rapid, especially at the time when they are about half their final size. Leaflets of *Amherstia nobilis* can increase in length by 18 mm per day, and petioles by 41 mm per day (Schimper and von Faber 1935). Leaf growth in this rapidly flushing species is sometimes completed in less than

Fig. 5.11 Variation in the size and shape of fully grown leaves of *Triplochiton scleroxylon*. The 5 smaller leaves of clone 8 035 were produced by weak shoots in unfavourable conditions for growth; the 2 large leaves on vigorous shoots in a favourable environment. Note the characteristic secondary lobing in clone 8 036, and the drip-tips.

two weeks, although it takes another fortnight for the leaves to spread out horizontally. The rachis of the large fern *Angiopteris evecta* can even extend at 90 mm per day, and Coster (1927) found that leaf growth rates of these two species were lower in the day than at night when the weather was sunny, but not on a cloudy day. In the great majority of species, leaf growth is determinate, so that the duration of its period of expansion is limited. Nevertheless, in some species with large, compound leaves, each one may take several weeks to complete its growth. In a few instances, such as *Guarea guidonia* and *Chisocheton* spp., the leaf-tips remain meristematic, producing new leaflets continuously or periodically over a longer period of time (Steingraeber and Fisher 1986).

Leaf growth usually involves both cell division and cell enlargement, and the final size reached, and sometimes also the leaf shape, can be affected by a range of external, internal and genetic factors (Fig. 5.11). A striking example of unequal growth of adjacent leaves is provided by plants showing strong **anisophylly**, for example the tropical American *Columnea sanguinea*, the Malaysian *Anisophyllea trapezoidalis* and the West African *A. laurina*. The branches have two rows of large leaves below, and two of very small leaves above, while on vertically growing main stems all leaves are large. A possible hypothesis might be that the horizontal position of the branch in relation to the gravitational field imposes a **gravimorphic** effect, so that either inhibition of growth occurs in leaves inserted on the upper side, or the two types of leaf arise from primordia that contain very different numbers of cells. Less pronounced anisophylly is found in other genera, for example, *Alstonia* and *Gmelina*.

Water stress is one of the factors likely to reduce leaf growth, possibly through the production of large amounts of ABA which may close stomata. In cocoa, for instance, when the available water content of the soil had decreased by a third, leaf growth had slowed considerably (Lemée 1956). By the time the moisture supply had decreased by a further third, expansion had stopped entirely, so that leaves only reached a length of 25–60 mm, while the control leaves on plants in moist soil were twice as long and still growing.

Shading commonly influences leaf development, generally increasing the rates of growth and final sizes, and sometimes modifying leaf shape. In cocoa and coffee, for instance, which are small trees of the forest understorey, the leaves that expand in full sunlight are often smaller and yellowish-green, while those growing in shade tend to become larger and darker green. Interestingly, it has been found that cocoa leaves in full light grow considerably larger if the lower mature leaves on the same shoot have been removed (Jan Krekule, unpublished data). The implication is that the effects of light intensity and quality may be indirect, perhaps operating via water stress. Even in rigorously controlled environments, it is quite difficult to distinguish between genuine control by photon flux density, and secondary effects such as water stress or leaf temperature changes, or tertiary responses such as the closure of stomates. In shading trials in the field, differing air and soil temperatures also have to be taken into account, together with variation in atmospheric saturation, vapour pressure deficits and any root competition from shade trees.

Temperature and photoperiod can affect leaf growth; thus seedlings of *Triplochiton scleroxylon* grown under 11 h days + 20 °C nights produced leaves that were 35–40 per cent shorter than on plants receiving either 14½ days, or 30 °C nights, or both (Fig. 5.6B). *Ceiba pentandra* and *Gmelina arborea* also showed an interaction between these two factors, while in *Chlorophora excelsa* leaves were 50 per cent longer when grown under long- rather than short-days (Longman 1978). The effects on leaf area were likely to have been even greater than is suggested by these assessment of linear dimensions. Leaves of a paler, yellowish-green colour tended to occur when *Triplochiton scleroxylon, Terminalia ivorensis* and *T. superba* plants were grown in cool day/hot night regimes, in comparison with other combinations of fluctuating or constant temperatures.

There are likely to be many other influences upon the development of a leaf, including such factors as the availability of nitrogen and mineral nutrients, and the influence of herbivore damage to it (and elsewhere in the same or neighbouring trees). Changes in leaves which occur with the increasing physiological age of the tree are considered in section 5.8, while the longevity of individual leaves is discussed in the section which now follows.

5.5 Leaf senescence and abscission

A leaf typically reaches a peak of photosynthetic capacity around the time that expansion ceases, thereafter showing a gradual decline. Sooner or later a point is reached when it rather suddenly begins to turn yellow, brown, or less commonly red, and is then actively shed from the tree by completion of a separation layer in the abscission zone at its base. This change of colour is termed senescence, and it appears to be mediated by alterations to the balance of several endogenous growth substances. It is the terminal irreversible phase in the functioning of the leaf (Addicott 1982), which signals the wholesale breakdown of chlorophyll, ribonucleic acid and protein, and the rapid translocation out into the stem of some but not all of its organic and inorganic nutrients.

Abscission normally follows quickly, and the protective layer of the leaf-scar becomes an effective barrier to invasion by pathogens, except for instance in cocoa where it may become a centre for *Phytopthora* infection. Lack of leaf abscission, with shrivelled leaves still retained on the stem, is often a sign that sudden stress has prevented the usual sequence of changes; retention of dead leaf bases is, however, a normal feature of some plants, including palms and tree-ferns. In *Cupressus* and *Pinus*, whole leafy branchlets are shed, rather than individual leaves, while in *Taxodium* leaves are shed from long-shoots, but the entire short-shoot is abscised.

A very wide range of factors appear to promote leaf-fall in trees, including lowered light intensity, changed temperature and photoperiod, mineral nutrient deficiency and water stress. Older leaves are much more likely to be abscised than those which have recently expanded, and it is probable that there are also interactive effects between different organs. For instance, container-grown cocoa plants subjected to drier conditions abscised a higher proportion of their leaves, and there was also evidence for an influence of the shoot/root ratio (Sale 1970; Borchert 1978). Correlative effects were also suggested in an experiment with *Hildegardia barteri* seedlings, in which, unexpectedly, more leaves were abscised under long-days than in short-days. In the former treatment, new leaves were being produced and expanded throughout the experiment, whereas the terminal buds on plants in short photoperiods were dormant for a long time. Table 5.3 shows what appears to be a direct, additive effect of short-days and hot nights in enhancing leaf abscission in *Bombax buonopozense*. Two stages of leaf loss were detected in *Plumeria acuminata*, both of which were inhibited by light interruptions during the dark period (Murashige 1966).

Table 5.3 Interaction of day-length and night temperature on leaf abscission in *Bombax buonopozense* seedlings. Mean number of leaves lost from each plant in a period of 4 months

	Long-days	Short-days
26 °C nights	0.6	1.5
36 °C nights	3.0	6.0

An interesting example is that of a disease of coffee caused by the fungus *Omphalia flavida*, which leads to premature leaf-fall and greatly reduced yields. Abscission appears to be stimulated not so much by leaf damage *per se* as by disturbance of the balance of hormones. Once the flow of auxin from leaf to stem is reduced, the separation layer at the base of the petiole completes its development, and the leaf is abscised within a week (Sequeira and Steeves 1954).

Leaf-fall occurs all the year round in tropical forests, but peaks and troughs are usually found. The classical studies of Bray and Gorham (1964) on seasonal production of litter (about two-thirds of which was foliage) showed that in the rain forests of Colombia and Ghana, for example, most litter accumulated in March and least in July. In evergreen seasonal forests, it appears that leaf-fall may peak in the first half of the dry season, as some of the trees become completely leafless; evergreen specimens may also lose a proportion of their leaves at this time. Fluctuations in leaf-fall have many ecological implications, for instance regarding the nutrient and water status of the soil, the photon flux densities reaching the lower tree layer, and the amount of food available to herbivores. Every species does not behave in the same fashion of course, individual specimens in a population are often not synchronised, and conditions may vary from year to year. Nevertheless, regular annual leaf-fall has been recorded, for instance, around February for *Instia palembanica* in Malaysia. In the life of an individual tree, the partial or complete loss of its old foliage represents an important change in photosynthetic, transpiring and respiring tissue (see, e.g., Janzen and Wilson 1974), except when leaves are produced and lost steadily throughout the year.

Rather little attention has been paid until recently to the important topic of the longevity of tropical leaves (Bentley 1979; Chabot and Hicks 1982). There is no difficulty in knowing the age of a leaf on the majority of temperate zone trees: if they are deciduous species, then any leaves must be of the current year; while in many, though not all evergreens, there is distinctly 'articulate growth', such that the leaves produced in successive years are in separate 'cohorts'. In the tropics, on the other hand, it is usually impossible to know the age of a leaf unless trees are observed regularly, preferably with repeated records made on tagged leaves. The minimum life-span for undamaged leaves is likely to prove to be about three months, as recorded for the woody climber *Grewia carpinifolia* in Ghana (Swaine *et al.* 1984). Although the majority of leaves probably do not last more than about fifteen months, there are some well-documented cases, such as the understorey trees *Drypetes parvifolia*, *Vepris heterophylla*, and canopy species *Diospyros abyssinica* and *D. mespiliformis*, and conifers of montane forests, in which the life-span is at least 2–3 years (Swaine *et al.* 1984; Grubb 1977).

There is a considerable amount of confusion in the literature over the distinction between the evergreen and deciduous habits in tropical forests. Basically, the question of whether a forest, tree or branch is leafy or leafless is a function of longevity and the relative timing of bud-break and leaf abscission, perhaps modifed by herbivory and other damage. Because all shoots on a tree (Fig. 5.12), all individuals of a species, and all species in a forest are often not synchronised, recognising firm 'leafiness' categories becomes increasingly difficult the larger the unit. Indeed, even within a single shoot, the older leaves may be shed before those more recently expanded, while in compound leaves some of the leaflets may be abscised and others retained.

Taking these points into consideration, the following four broad categories are proposed as useful groupings that are based on recognisably different growth strategies, and can be applied at any level:

A. **Periodic growth (deciduous):**
Leaf-fall occurs ***well before*** *bud-break; life-span about 4–11 months*
In this case, the branch, whole tree or forest is leafless (or

Fig. 5.12 Young *Ceiba pentandra* in S Ghana (early February), showing variable phenology. Lower left – flushing leaves; centre – older leaves; top right – leafless branches with ripening and shedding fruits.

nearly so) for a definite interval, varying from a few weeks to several months. Even in rain forest receiving a high rainfall evenly spread throughout the year, canopy trees of *Cordia alliodora* may for instance remain leafless for weeks in the middle of the rainy season in Costa Rica (Janzen 1975). In seasonal forests in West Africa, some specimens of *Terminalia ivorensis* can remain leafless for more than 6 months, mainly during the drier months. In this type, leaf-fall and bud-break are apparently not directly connected with each other.

B. **Periodic growth (leaf-exchanging):**
Leaf-fall ***associated with*** *bud-break; life-span often about 12 (or 6) months*
Here the flushing of new leaves starts around the time at which the majority of the old ones are abscised, within about a week either way. This habit is clearly seen in *Terminalia catappa*, where the old leaves turn red and fall and are simultaneously replaced by new (Fig. 5.13). In this species, the process often occurs twice a year (Koriba 1958), while in *Entandrophragma angolense, Dillenia indica, Ficus variegata* and *Parkia roxburghii*, for example, it happens once. In upland forest in Costa Rica, *Quercus* spp. can be seen to shed all the old leaves, and then within a few days to be expanding many new red leaves. It is likely that it is the

Fig. 5.13 Leaf-exchanging habit in *Terminalia catappa*. The old leaves are senescing at the same time as the new ones flush. The separation layer of a leaf that has already been abscised can be seen (bottom, centre).

senescence of the old leaves which provides the physiological signal for bud-break, placing type B responses in a different category from either deciduous or evergreen. The term 'leaf-exchanging' should be distinguished from the somewhat confusing expression 'leaf change' (*Laubwechsel*), which tends to incorporate elements of flushing, leaf growth and abscission. 'Semi-deciduous' is unsuitable because it is often used to imply that a substantial number of trees in a forest are deciduous.

C. **Periodic growth (evergreen):**
Leaf-fall completed ***well after*** *bud-break; life-span about 7–15 months or more*
This is a truly evergreen category, which is characteristic of woody plants of the lower and middle tree layers, and some of the upper tree layer. Many dipterocarps are periodic/evergreen, and other examples include *Clusia rosea, Fagraea fragrans, Mangifera indica* and *Pinus* spp. As in type A, there may be no direct connection between leaf-fall and bud-break, but it is possible that the expansion of new leaves might lead to senescence of the old through a hormonal control system, by competition for nutrients, or because of shading.

D. **Continuous growth (evergreen):**
Continual *formation and loss of leaves; life-span variable from about 3–15 months*
This is a second evergreen class, and is characteristic of some palms, conifers and tree-ferns, and is found in seedlings of many dicotyledonous trees. Examples of older trees include *Dillenia suffruticosa*, *Trema guineensis*, *T. micranthum*, and the shrub *Hamelia patens*. In this type there is, of course, no bud-break, but the rates of leaf initiation, enlargement and abscission may each vary considerably, according to changes in the environment, or to external and internal competitive effects. The 'leafiness' may thus fluctuate, although this was not found to be the case in the red mangrove, *Rhizophora mangle* (Gill and Tomlinson 1971).

The presence of type A trees even in the most uniformly moist of rain forests is an important point that is often overlooked. It provides a reminder that the microclimate in which leaves in the euphotic layer develop, particularly on the exposed crowns of emergents, is quite unlike the ever-constant, equable conditions of journalistic fables, and can be quite stressful (see sections 3.1–3.3). In evergreen seasonal forests, one-third of the tree species may show the deciduous habit (Taylor 1960), and the

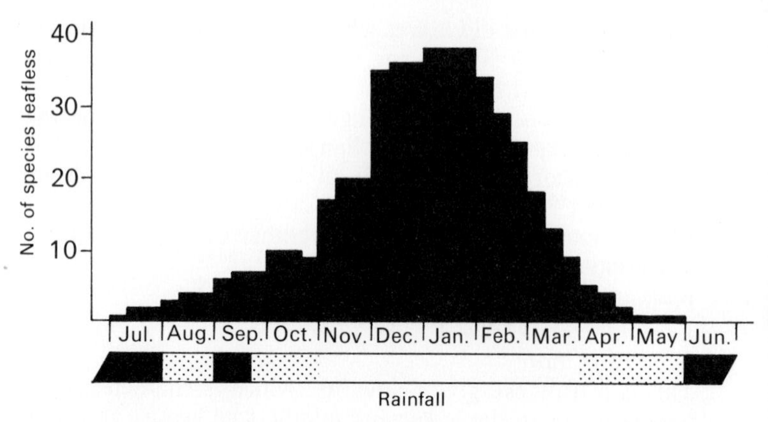

Fig. 5.14 Seasonality of the leafless condition in 158 forest tree species in Ghana. (After Taylor 1960.)

timing of the leafless period in Ghanaian trees closely followed the rainfall patterns. Figure 5.14 shows that there were some leafless trees to be found in every month except June, but the proportion of trees without leaves increased to a peak at the height of the dry season in December and January. The percentage declines as flushing occurs in the majority of species in February and March, and falls to a very low level before the onset of the rainy season.

In evergreen seasonal forests, as in other drier formations, there seems little doubt that the chief adaptive value of the deciduous habit lies in the avoidance of severe water stress in the middle part of the dry season. Yet it should not be forgotten that many other trees operating on type B and C strategies endure the same dry seasons in a leafy state, just as various evergreen trees survive the cold winters at higher latitudes, while other species do so leafless. Not just one, but several kinds of developmental strategy commonly evolve in response to environmental or biotic problems. Moreover, drought is clearly not the only stress involved because some species lose their leaves in the rainy season, perhaps because light intensities are lower than in the sunnier dry season. It should also be borne in mind that, as Janzen and Wilson (1974) have pointed out, there may be an extra 'cost' attached to being leafless in the tropics, since stored carbohydrate can be reduced by as much as 50 per cent because of high respiration rates in the twigs, boles and roots.

Although the usefulness of the underlying feature of 'leafiness' is beyond doubt, it has been lessened because of considerable variations in usage and precise meaning of the terms used. The frequency with which the above four categories of leaf and bud behaviour occur provides a basis for improving the physiognomic classification of tropical forests (see section 6.5). A useful distinction has been made, for instance, between facultatively deciduous trees, such as *Ochroma pyramidalis* and *Tectona grandis*; and obligately deciduous species, such as *Ceiba pentandra, Cordia alliodora* and *Enterolobium cyclocarpum* (Medina 1983).

Experimental studies on the relationships between leaf abscission and the growth and development of other organs may also be stimulated if these classes are used, for example the effects of defoliation on bud-break in type B and C shoots. Although not discussed in detail in this chapter, one of the most important consequences of all the different strategies of leaf development and maintenance will be the overall influence they have upon dry weight accumulation in the tree and forest as a whole, and on the yields that may be obtained from tree crops. However, some assimilates will be used up in producing the organs discussed in the remaining sections.

5.6 Cambial activity

A further dimension to growth and development is the increase in thickness of organs through the activity of the vascular cambium. Indeed, the secondary tissues produced by this meristem constitute much the greater part of the plant body in large perennials, except for palms and bamboos. In time, individual tropical trees can become extremely large, with extensive crowns and massive root systems, and diameters at the base of the trunk (exclusive of any buttresses) exceeding 2 m. Details of some big specimens are given in Richards (1952), including an *Entandrophragma cylindricum* in Nigeria which measured over 5.4 m diameter at the base. There is also a record of a 4.1 m diameter *Bertholletia* spp. in Brazil, and a 3.8 m *Balanocarpus heimii* in Malaysia. Such trees may have total dry weights of the order of 100 t, and (although not as large as some of the giant redwoods, eucalypts and *Agathis*), they rank amongst the largest living organisms in the world.

Yearly increases in main stem diameter vary greatly in different species, and from one individual to the other. As the trunk grows in size, the cross-sectional area of new tissue required for a given diameter increment gets larger. Thus, for example, what is produced by a 0.1 m diameter tree when it grows 10 mm would only increase a 1 m diameter tree by about 1 mm. In middle age, trees in natural forest that have their leaves mostly in the euphotic layer may add around 5–20 mm, but in very old emergents the increment is generally lower, sometimes below 1 mm (Jones 1956). In an analysis of 120 000 trees in an Amazonian rain forest, it was predicted from the stand structure that the average tree in the diameter class 0.25–0.35 m would be growing at about 8 mm a year; for larger trees the increment would decline below 4 mm (Heinsdijk and Miranda Bastos 1963). In a wet tropical forest in Costa Rica, analysis of 13-year increments of 45 tree and one liane species by growth simulation showed a range of median annual diameter increases between 0.3 mm and 13.4 mm (Lieberman *et al.* 1985).

Young saplings and pole-stage trees often make slow radial growth, suppressed by shade cast by the canopy trees, and competition from their root systems. In gaps and along forest margins, however, the same species may be capable of rapid growth, while colonisers may be even faster growing. In plantation conditions, too, substantial annual diameter increments can be sustained: for example approximately 45 year old *Triplochiton scleroxylon* in line enrichment plantings at Mabak, Cameroon, were still producing 12 mm (Leakey and Grison 1985). Exceptionally in young, dominant individuals, values of 20–60 mm can occur, and 90 mm has even been recorded in *Ochroma lagopus*.

Although there is little problem in relating approximate average diameter increments with age in fast-growing plantations, accurate measurement of the rate and duration of radial growth is usually difficult in tropical woody plants. Increase in thickness is generally the result of activity by two different lateral meristems, the vascular cambium and the phellogen (cork cambium), both of which may be cutting off cells towards the outside of the trunk and towards the inside. Moreover, successive girth or diameter readings do not necessarily show the combined activity of the two cambia, even if a standard measuring position is marked on the trunks (generally 1.3 m, on the upper side of the tree if there is an appreciable slope).

In the first place, this is because there may be irregularities present such as ridges or spines, branch and leaf bases or scars, or major convolutions in outline (Fig. 5.15). Secondly, bark may be abscised in various amounts and fashions (see Addicott 1982), sometimes in whole sheets, as in *Eucalyptus* spp. and *Tristania* spp. of the Myrtaceae. Thirdly, tree trunks regularly shrink appreciably during a hot, sunny day, because the xylem sap is under considerable negative pressure. Accurate measurement therefore requires continuously recording dendrometers, unless all the manual recording can be completed by 0600 h! The existence of tension in the conducting elements of the xylem in

Fig. 5.15 Cross-section of *Salacia pyriformis*, a woody climber from W Africa. The false rings do not denote annual xylem increments, but consist of included phloem produced during abnormal secondary thickening.

the middle of the day can be simply demonstrated in *Dillenia* spp. and some lianes if the crown has been in sunshine for some time. If a clean cut is made into the stem with a machete, a distinct sound can often be heard as air is drawn in.

In many cases, interest centres on the major component, the production of xylem by the vascular cambium. Increment cores or anatomical examination are needed to provide accurate information, and a further problem arises in natural forest from the frequent lack of clear and reliable annual growth rings in many tree species (Fig. 5.15). The problem of knowing how old tropical trees are introduces inaccuracies, not just for calculating cambial growth rates, but also for every other aspect of the analysis of growth and stand structure in tropical forests.

There is some prospect of newer methods being developed, such as modelling and detailed examination of the transverse surface for small changes in chemical or physical structure. For instance, the $^{14}C/^{12}C$ ratio in carbon dioxide changed during the last 30 years, because of the atmospheric testing of nuclear weapons in the late 1950s and early 1960s, before this was banned by international agreement. This left a characteristic 'thumbprint' in the permanent structural cell components of all woody plants, which varies only slightly according to latitude (Bormann and Berlyn 1981). For many purposes, simpler methods may be more appropriate, such as the marking of the xylem at a particular point by dyes or damage with pins or nails, or by opening a small annual 'window' (Shiokura 1980; Mariaux 1981). The living cells in the wood often remain free of starch for a considerable time after they have been formed, which could be used to determine the increment in numbers (Fink 1982).

In order to understand more about the physiology of cambial activity, it is probably best to concentrate on species with clear-cut, annual rings. There are more of these than is generally realised, including *Homalium tomentosum, Pterocarpus indicus* and *Tectona grandis* in South-East Asia; *Bursera simarouba, Goethalsia meiantha, Swietenia mahogani* and *Taxodium distichum* in Central America; *Ceiba pentandra, Chlorophora excelsa, Entandrophragma* spp., and *Terminalia* spp. in West Africa (Fahn *et al.* in Bormann and Berlyn 1981).

Most of these are deciduous or leaf-exchanging species (types A and B of section 5.5), and the former group were shown many decades ago to exhibit dormancy of the cambium. By a combination of careful measurements, experiments and anatomical studies, Coster (1927–28) demonstrated the cessation of cambial activity at the beginning of the leafless period, with divisions occurring again when flushing took place. He also showed that the stimulus for renewed cambial activity in leafless plants must arise from the expanding buds, since the cambium remained dormant if they were removed, or if they were isolated from the portion of cambium under study by complete ringing. The influence of the expanding buds was not exerted through photosynthesis, for the cambium was still activated if the plants were kept in the dark, so Coster proposed a hormonal explanation at a time when plant growth substances had only recently been discovered.

Thus it is likely that auxin, moving basipetally in the phloem from expanding buds, is primarily responsible for initiating cambial activity. Other hormones are probably involved as well, since for instance the cambium in dormant, leafless teak shoots did not respond to IAA unless bud dormancy was artificially broken with ethylene chlorhydrin (Reinders-Gouwentak 1965). After leaf expansion has been completed, it is generally considered that there may be sufficient auxin produced by the mature leaves for cambial activity to be maintained, but when they all senesce the supply dwindles and the cambium becomes dormant. Indeed, Coster was able to cause the cessation of cambial activity by ringing the stem between the mature leaves and the part of the cambium being studied.

Conversely, there is recent evidence that the living cells of the wood may influence primary shoot growth. The vertical parenchyma cells, especially those associated with the vessels (Fig. 5.16), show high levels of acid phosphatase enzymes, particularly in the pits linking the two types of cells. It seems likely that sugars are being secreted into the vessels, as occurs in the temperate zone sugar maple tree, but for much longer periods, and in more species in the tropics (Fink 1982).

The timing of the periods of cambial activity and rest is thus likely to be controlled indirectly by the same external and internal factors which determine bud-break (section 5.1) and leaf-fall (section 5.5). One exception is probably provided by trees that flower during the leafless period, where opening floral buds and developing fruits could provide the stimulus for an earlier resumption of cambial activity. This is one of the many possible reason for 'false' rings in the xylem, which mean that not every deciduous tree shows clear-cut annual growth patterns in its wood.

In the leaf-exchanging trees of type B, which are leafless for at most a few days, the cambium may not become fully dormant. Indistinct rings were formed in *Khaya grandifoliola*, but clear annual rings in *Dillenia indica* and *Ficus variegata*. In evergreen trees of type C, continuous cambial activity is probably the rule, although a few samples of cambial dormancy and even annual rings have been reported (Alvim 1964). More generally, they are indistinct, irregular in occurrence or absent, especially when seasonality of climate is slight. The clear-cut growth rings in *Avicennia germinans* consist of alternating bands of wood and softer phloem-containing tissue, and do not indicate age. The most vigorous shoots can contain up to 5 rings, all of uniform width (Gill 1971). In evergreen trees of type D, including young saplings that are continuously expanding new leaves, the cambium is almost certainly active throughout the year.

Whether cambial activity is continuous or periodic, there are often large fluctuations in the rate at which cells are formed and differentiated. As generally recognised, there is a clear positive correlation between the rate of diameter increment and the amount of height growth being made by a tree, and there is often an even closer link with its crown dimensions. Cambial activity may in some cases be most rapid while shoots are extending, but in cocoa it was slow while flushing proceeded, and reached a peak during the rainy season (Alvim 1964). Similar fluctuation has been recorded for other tropical tree species (Hopkins 1970).

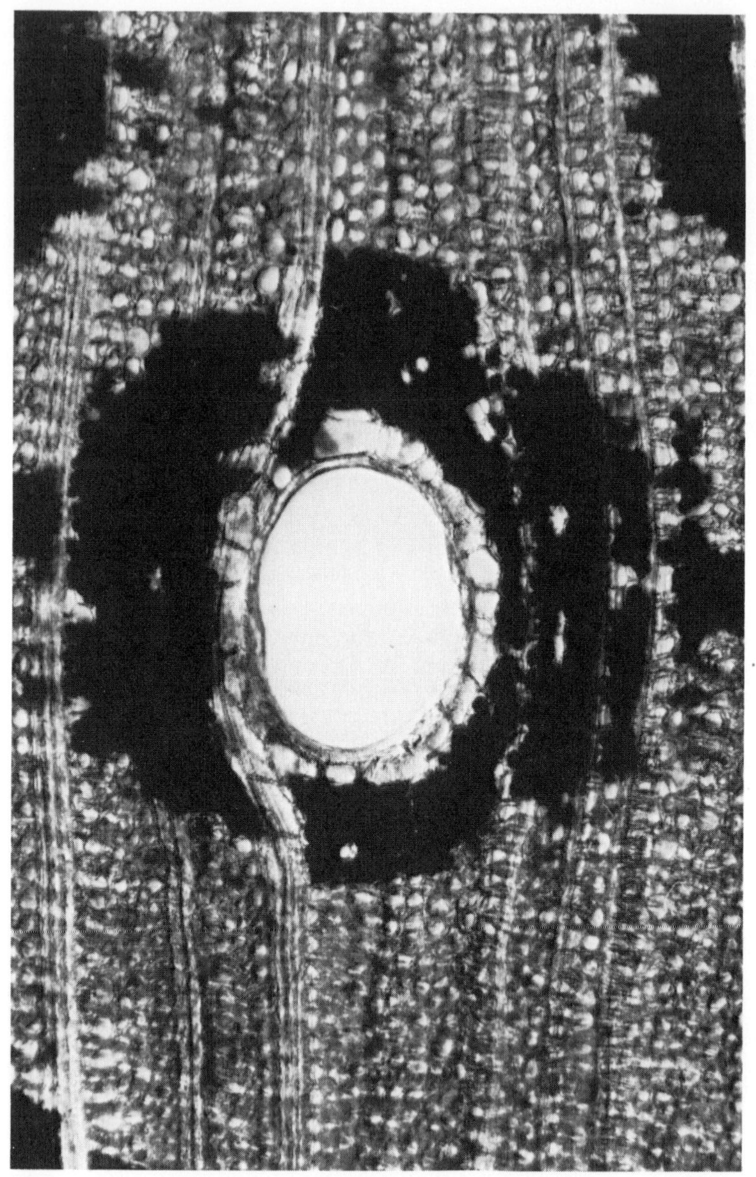

Fig. 5.16 Transverse section of the wood of *Enterolobium cyclocarpum*, showing plentiful starch (black areas) in the vertical parenchyma surrounding the vessel, except for the cells in contact with it. These and the rays are completely starch-free. (From Fink 1982.)

The factors which control the rate of cambial activity have been little studied experimentally, in spite of their obvious importance to forestry. Some external influences may be expected to act directly on the cambial initials or the differentiating cells, as for example temperature and moisture stress, or the effect of gravity in branches and leaning main stems. Others, for example photoperiod and mineral nutrition, probably affect the cambial zone indirectly, perhaps by modifying growth or internal control systems, such as the partitioning of assimilates between various active meristems or into storage sites. High temperatures between 35 °C and 40 °C, which reduced the rate of shoot extension in *Ceiba pentandra* compared with somewhat cooler conditions, also significantly depressed diameter increments (Kwakwa 1964). Water stress was implicated in a study of cocoa, in which plants supplied with the equivalent of the annual rainfall (1, 830 mm) spread evenly over the whole year made about 30 per cent more diameter growth than those receiving natural precipitation (Murray 1966).

Considerable genetic differences within species in their diameter increment have frequently been reported between provenances of tropical forest trees. Pronounced clonal variation also occurs: for example, both the radial increment of xylem and the number of cells produced radially was more than twice as great in one clone of *Terminalia superba* than in another (Longman *et al.* 1979). In terms of cross-sectional area (basal area) or volume production of wood, such differences become even larger, since the calculations involve squaring the data.

Significant clonal differences also occurred in *T. superba* in the radial dimensions of vessels and of fibres + parenchyma cells, which may be expected to alter the wood properties. These will clearly be affected as well by the proportion of the different types of cell that are produced. However, although the patterns of different tissues have been extensively described, and are often of value in identification of timbers, there is as yet little understanding of what determines the type and size of cell produced. How, for instance, do the narrow, tangential bands of parenchyma and tracheids which mark cessation of cambial activity at a growth ring differ from those which are often formed during continuing wood production? Scanning electron microscopy in association with detailed studies on clonal material growing in controlled environments might clarify this. Moreover, there is still much to be done in relating the structure of wood to its properties, and here the work of botanists and those in the timber trade may be assisted and integrated by a useful handbook by Chudnoff (1984), which covers the extremely confusing area of local and botanical names for 374 different timbers, as well as many aspects of their utilisation.

5.7 Root formation and development

The first roots to be formed in a young seedling are usually a group of first-order laterals that arise from the tap-root just below ground level. Like most roots, these do not grow vertically downwards but elongate more or less horizontally. If the tip of the primary root of young cocoa seedlings is decapitated, some of the laterals nearest to the cut curve downwards and become positively geotropic (Dyanat-Nejad 1970). Thus even at this very early stage, it is clear that there are dominance relationships within the root system, and these presumably lie behind the various characteristic organisational patterns found in older trees (see Fig. 4.11). Because of the difficulties involved in following individual subterranean roots, the majority of work on branching has been done using aerial roots. Unless damaged, these typically show little or no branching while in the air, but lateral roots start to appear if the tip makes contact with the ground. In *Cissus*, branching is induced if the root is placed in water, but in *Rhizophora* darkness appears to be the chief stimulus (Gill and Tomlinson 1975).

The initiation of aerial roots takes place within the shoot system, providing an opportunity to examine the formation of new root apices above-ground (see Fig. 4.8). Trees with stilt-roots, stranglers and epiphytes may all provide potential material, while climbers such as *Cissus aralioides* and *Urera* spp. actually spread vegetatively through the formation of adventitious roots (Hall and Swaine 1981). *Anthonotha macrophylla, Scaphopetalum amoenum* and *Sleotiopsis usambarensis* form new roots during the natural layering of their drooping branches; while conversely the rarer case of the formation of shoot apices on root tissue could be studied in plants forming root suckers, such as *Bumelia reclinata, Cedrela serrata, Chlorophora excelsa, Cordia alliodora, Drypetes diversifolia, Melia azedarach, Millettia thonningii* and *Trema* spp.

Despite these and other instances, vegetative spread is thought to be relatively uncommon amongst the woody plants of the tropical forests. However, the capacity to produce roots in stem cuttings appears to be widespread (Leakey *et al.* 1982*a*); particularly if the material originates from young trees (Fig. 5.17), and so perhaps natural vegetative reproduction may turn out to be commoner than expected. Since many commercially important plantation species can be rooted from stem cuttings, this has a

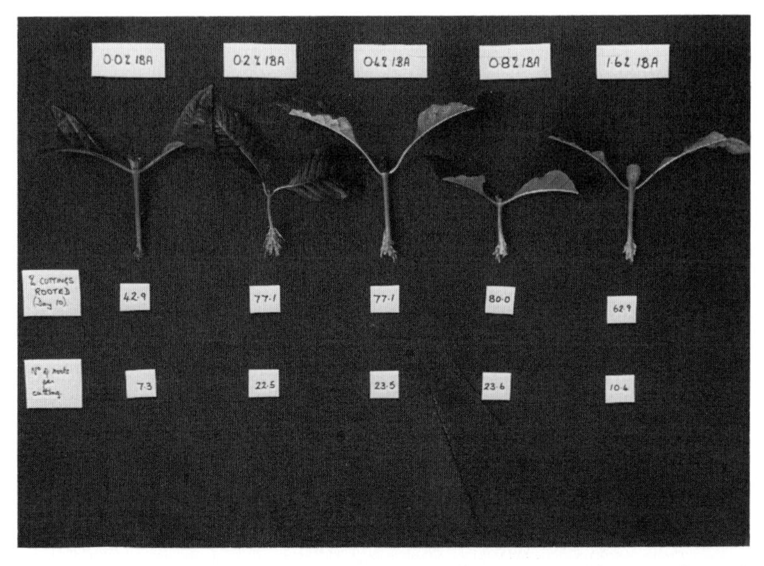

Fig. 5.17 Effect of auxin concentration on rooting and root growth of leafy, single-node cuttings of *Nauclea diderrichii* under automatic mist. (IBA = indolebutyric acid.)

number of important implications for research, tree improvement and clonal forestry (see section 7.4).

It also provides favourable circumstances for studying the formation and outgrowth of roots, since large numbers of cuttings can be set and removed at intervals from relatively uniform propagation beds. Few anatomical investigations have been done on tropical trees, but there is a growing literature on environmental, hormonal and other effects on root initiation. For instance, rooting of cuttings of *Triplochiton scleroxylon* under automatic mist was promoted when the temperature of the propagation medium was raised to 25 °–30 °C, compared with 20 °C (Fig. 5.18), and these conditions appear to be suitable for many other genera.

There are often substantial differences in rooting ability between clones; and also amongst cuttings that are of different sizes, originate from different parts of a tree, or from trees pretreated in various ways (Leakey *et al*. 1982*a*; Leakey 1983; Lo 1985; Leakey and Mohammed 1985). The addition of auxins may increase the rate of rooting, and the number of root initials formed, but when supra-optimal can inhibit root elongation (Fig. 5.18). Rooting with and without auxin in the medium has even

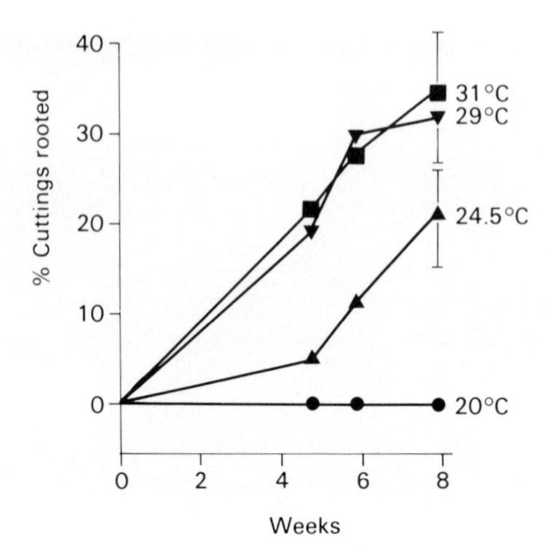

Fig. 5.18 Effect of propagating bed temperature on the rooting of leafy, single-node cuttings of *Triplochiton scleroxylon* under automatic mist and without auxin. (From Leakey *et al.* 1982*a*.)

been achieved with teak in aseptic culture, using small shoots produced by meristem proliferation (Houaye 1983).

The daily course of root elongation was followed in aerial roots of a large liane, *Cissus adnata*, whose foliage was high in the forest canopy (Coster 1927). The roots showed little or no growth when the sun was shining or when a drying wind blew during the night, but immediately after rain the rate of elongation rose temporarily to the very high value of 30 mm/h or more. Coster showed that the chief factor involved was the water status of the plant, and found that the roots could actually shrink by as much as 7 per cent in length when subjected to drying. In contrast, the aerial roots of *C. sicyoides* grew at a constant 4 mm/h throughout the night and a sunny day.

Seasonal changes in the rate of root growth no doubt occur widely, but have been little studied. Nor is it clear whether there is any general cessation of root activity at certain times of year, as there usually is in the shoot system. Only in extreme cases does temperature appear to stop elongation: *Citrus* roots for instance grow only above 10 °C. However, it may well be that root and shoot growth are linked by internal control systems as well as being influenced by their separate external environments. For instance, in controlled environments the root and shoot growth of cocoa are both rhythmic, with peaks and troughs of activity alternating. The rapidly expanding leaves of the flush apparently inhibit the roots, because if the foliage was all removed immediately after unfolding, the roots showed a high constant growth rate during this period (Sleigh *et al.* 1984).

Coster (1932) grew about 70 tropical (mainly woody) species at a site in Java which had a deep, permeable soil and regular rainfall. After 6 months, he dug out the seedling root systems carefully and found that in general the main root was longer than the main stem, and the spread of the horizontal surface roots was greater than that of the crown. Average root elongation rates were over 20 mm per day for the most rapidly growing trees, which is faster than that found in most temperate trees (Lyr and Hoffmann 1967). The total length of the main root plus the primary or first-order lateral roots had actually exceeded 30 m in *Melia azedarach* and *Sesbania sesban*, though others had grown much more slowly.

Just how extensive root systems can become is vividly illustrated by the fact that coffee plants can produce roots totalling 25 km in length by the time they are three years old. At La Selva, Costa Rica, it was estimated that more than 250 g m^{-2} dry weight of fine roots was produced in the first year by woody regrowth (Raich 1980). This was roughly comparable to that present in the 5-year-old secondary forest that had been cleared, indicating a very rapid restoration of the nutrient uptake capacity. Fine roots often have a short life-span: it took much longer for the new structural root systems to develop. A major problem that hampers studies of the dynamics and turnover of root systems is the long, hard and tedious work involved in digging out, separating and assessing the different types of root (see section 4.2). It is seldom possible to attempt sufficient replicates to assign variation reliably to any factor, but one exception is provided by a study of the maximum rooting depth in tea. This indicates a significant difference between 8 clones, the deepest growing of which had reached 2½ times the depth of the shallowest in the same time (Nagarajah and Ratnasuriya 1981).

Cambial activity in roots is another area which has been little studied (Fig. 5.19). When aerial roots of *Rhizophora mangle* reach the ground, their extraordinary rapid extension rates slow, and they cease to possess an elongating zone averaging 12 mm in length (Gill and Tomlinson 1975). Soon afterwards cambial activity becomes very marked and the roots thicken. An unusual feature is the joining together through anastamoses and secondary thickening of two roots of the same tree, termed self-grafting by Ng (1975), and noted in several species in addition to *Ficus* spp., including *Shorea leprosula* and *Swietenia macrophylla*. Grafting between different trees was found in *Gmelina arborea*

Fig. 5.19 Root system of *Koompassia malaccensis* in a W Malaysian forest.

and *Shorea curtisii*, and Leroy Deval (1974) considers that dominant trees in stands of the gregarious species *Aucoumea klaineana* actually 'annex' the root-systems of dominated trees to which they are fused.

5.8 Physiological changes with age

One of the distinctive features of long-lived perennial plants is that they usually show variation in structure and function as they grow older. Some of the changes which occur are gradual, while others appear suddenly, but in either case shoots which are collected from the top of an emergent tree seldom resemble closely those produced by the same species as seedlings in the undergrowth (Roth 1984). In Fig. 5.20, the leaf of *Celtis mildbraedii* from the canopy is a quarter of the length and only 9 per cent of the area of that taken from the undergrowth, while in the liane *Hugonia platysepala* the canopy leaf is 40 per cent and 30 per cent of the shaded one. Sometimes leaf pairs of this kind are so unlike each other that most botanists would assume that they belonged to two different species or genera.

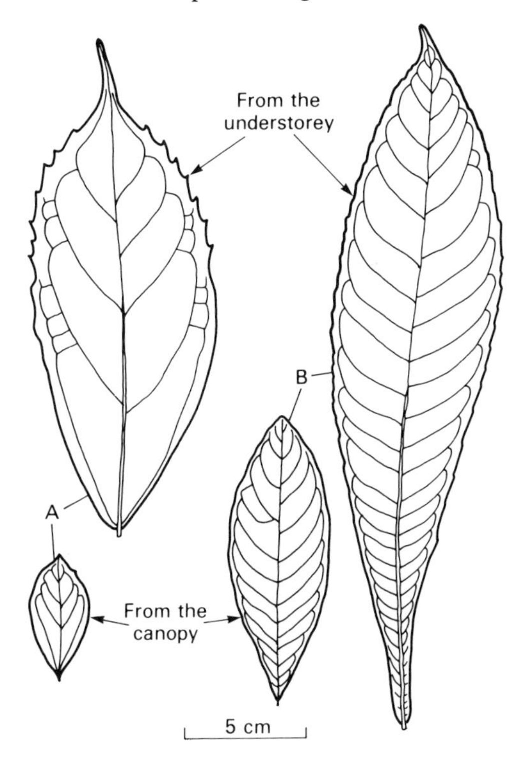

Fig. 5.20 Leaves from the canopy and the understorey of a Ghanaian forest:
(A) *Celtis mildbraedii*, a large tree; and (B) *Hugonia platysepala*, a liane. (From Hall and Swaine 1981.)

It is one thing to recognise that such within-tree variation occurs, but quite another to determine its cause. The leaves could be dissimilar for a large number of reasons that can be considered in four broad groupings:

Firstly, they have grown in very differing **environments**, so that factors such as light intensity and quality, temperature and moisture stress (see Chapter 3) may have modified them, directly or indirectly. Although it is difficult for logistic reasons to carry out experiments using emergents, it is possible in plots of saplings or batches of potted seedlings to investigate how much their leaf size and shape vary in response to treatment differences.

Secondly, they are normally bound to be of a different **genotype**. The extent of genetic variation is often great, especially in outbreeding species of forest trees that are 'undomesticated' (see section 7.4). It is well known that this can influence many aspects of growth, and thus affect leaf development directly and indirectly. With the increasing use of vegetative propagation in forest tree research and for clonal forestry, the strongly inherent characteristics of clones are likely to become as easily recognised as in *Hibiscus* or *Bougainvillea*, citrus or mango. For instance, potted plants of *Triplochiton scleroxylon* clone 8 036 could be readily distinguished from other clones by the pronounced secondary lobing of the leaves (see Fig. 5.11).

Thirdly, the leaves are borne on trees of a very different **size**, with much greater internal competition between potential sites for growth and storage in the big tree (i.e. many more 'sinks'), and longer distances for translocation. This question has been approached by experimental disbudding, using the short-lived woody plant cassava, with its relatively simple crown. This follows Leeuwenberg's model (see C in Fig. 4.12), with regular and repeated increase in the number of actively growing shoot apices, and associated decline in their vigour of growth. Since removal of all but one of the growing points more than doubled the rate of growth of a 'singled' apex by the end of the first week, and the leaves produced later were as much as 35 per cent longer than in unpruned controls, internal competition can indeed influence leaf size (Damptey 1964; Longman 1978). Differences in this category, which are due to **ageing** as a tree gets older, can typically be reversed by appropriate treatment.

Fourthly, the variation between the leaves may arise because of **phase-change**. These changes also occur with age, so that shoots originating from an older crown often exhibit more permanent differences from those taken from young seedlings (Hackett 1985; Longman 1986). The change from the 'juvenile' to the 'mature' phase is termed **maturation**, and the features are still exhibited when the mature tissue has been vegetatively propagated and the two phases are growing as similar-sized plants in the same environment. Ideally, investigation of phase-change should be done with material that is also of the same genotype, which is in fact possible if young coppice shoots can be obtained, as these are generally regarded as still being juvenile (see section 7.4). When such detailed investigations are not possible, a reasonable compromise might be to compare the crown foliage of several large trees with leaves taken from regeneration in a large gap or clearing nearby, where (away from the margins) the microclimate is relatively similar to that at the top of the canopy.

Differences that were clearly due to phase-change were found in similar-sized mature grafts and juvenile seedlings of *Cedrela odorata* that were grown in controlled environments under different photoperiods (Longman 1969). The rate of shoot elongation was much more strongly influenced by day-length in the mature plants, which also ceased extension for a time in both treatments. In contrast, the terminal buds of the juvenile trees only became dormant under short-days (Fig. 5.9). On the other hand, differences which are definitely not due to calendar age are shown in Fig. 5.21, since all the trees in the photograph are 5 years old. Because of the favourable external environment, one of the trees has far outgrown the others and has entered the category A phase (section 5.5) with periodic growth, a rapid flush of height growth and a clear-cut deciduous period. The small trees remained evergreen, and grew continuously or for long periods, though at a very slow rate.

In the uncontrolled setting of the tropical forest, it is generally difficult or impossible to separate environmental and genetic factors on the one hand from physiological changes with age (ageing and maturation) on the other. Differences of various kinds between the foliage of large and small trees are often reported, the leaves of the former generally being smaller (see Fig. 5.20), with drip-tips generally absent (see sections 3.3 and 4.2). Compound leaves often have fewer leaflets, as in the Meliaceae, for example. In some instances, the leaf shape, and sometimes the phyllotaxy, may alter strikingly with age; for example young seedlings of *Acacia mangium* have the normal pinnate leaves typical of the Leguminosae, but these quickly give way to phyllodes, with a few intervening nodes bearing transitional

Fig. 5.21 Eight seedlings of *Terminalia ivorensis*, all 5 years old, showing how powerfully environment can influence growth. Seven were very pot-bound, grew slowly and were evergreen. The eighth rooted through into the ground, cracking the pot, and was able to utilise the seepage of water and nutrients from the glasshouse. It grew very fast, and has already entered a deciduous stage.

structures. *Araucaria cunninghamii, Triphyophyllum peltatum* and some *Eucalyptus* spp. have at least three contrasting types of foliage.

As trees become larger, they often cease producing leaves continuously, and the leaf-exchanging or deciduous habits may appear. In genera which show periodic growth from the start, such as *Agathis, Barringtonia* and *Mangifera*, longer periods of terminal bud dormancy may be found. In addition, there may be a tendency for individuals of a species to become more synchronised in their development. For instance, the typically out-of-phase condition of a 2-3-year-old plantation of *Terminalia ivorensis* contrasts with the comparative regularity of a 20-year-old stand.

As a seedling grows into a large tree, the characteristic branching patterns also emerge (Tomlinson 1978; Hallé *et al.* 1978; Fisher 1978; see Fig. 5.22). Sometimes branches start to grow almost at once, while in other species there can be a period of a few months or years without any. In young saplings of some species in the Meliaceae, the rachis of large compound leaves borne on the main stem acts instead of a branch in supporting the photosynthetic surfaces. Palms that follow Corner's model (see Fig. 4.12), such as *Elaeis guineensis* and *Cocos nucifera*, produce very large fronds from their trunks, the only branches being reproductive and determinate.

Conversely, the terminal growing point of the main stem loses its meristematic nature and disappears in cocoa, signalling the production of a 'jorquette' of plagiotropically growing branches. Neither branches nor main axis are produced during the 'grass stage' of *Pinus merkusii*, most of the growth of the young tree going into the development of the root system. Only later do the main stem and branches develop as in other pines. The grass stage is thought to be an adaptation by which the terminal bud is protected from 'light' fires by the cluster of long needles: it occurs in this species in Vietnam, but not in Sumatra (Cooling 1968).

Some interesting theoretical modelling of branching is reviewed by Fisher (1984), but the amount of experimental work on branch development has not been extensive. The formation of the first primary branches was delayed by more than 3 months, and no second-order branches were formed, when *Rauvolfia vomitoria* seedlings were grown under 8 h rather than 12 h daylengths (Piringer *et al.* 1958). Applied gibberellin promoted the elongation of both orthotropic and plagiotropic branches in cocoa, but did not change their phyllotaxy or geotropic orientation (Greathouse and Laetsch 1973). *Triplochiton scleroxylon*

Fig. 5.22 Natural growth habit of *Alstonia* sp. at the Forest Research Institute, Kuala Lumpur, Malaysia. The terminal shoot apex is lost at intervals, and height growth is continued sympodially by orthotropic branches.

clones differed strongly in the number of branches per metre of main stem (Fig. 7.14), and in their thickness and persistence after they had become shaded by later growth (Leakey 1985*b*). Some clones showed a tendency to produce 'multistems' (several strong vertical shoots from near the ground). Decapitation studies with small potted plants have shown that clones vary in the number of buds which are released from apical dominance, and that mineral nutrition and gibberellins, for example, can affect the re-imposition of dominance by the uppermost lateral. Experiments with *Terminalia ivorensis* seedlings involving removal of terminal or lateral shoots showed that branches can affect the growth of the leading shoot, as well as *viceversa* (Damptey 1964).

Besides environmental and genetic effects, there appear to be various endogenous and ontogenetic factors at work: for instance, a clone of rubber trees usually formed the first branches after the terminal bud had flushed nine times (Hallé and Martin 1968). In some tropical trees there are two buds in each axil, and typically these behave differently. For example, the upper one in *Nauclea diderrichii* produces plagiotropic branches, while the accessory bud may produce an orthotropic shoot after damage or decapitation; and these differences persist in cuttings taken from the two types of shoot. There can be several accessory or serial buds in one axil (Fig. 5.23) and in coffee their dormancy can be broken by treatment with various growth regulators, including the cytokinin benzyladenine, the anti-auxin tri-iodobenzoic acid and compounds releasing the gaseous hormone ethylene in the tissues. The natural or induced sprouts from these accessory buds grow as orthotropic shoots, whereas the upper 'primary' bud produces plagiogropic, flowering shoots. However, the upper three buds can form flowers sequentially, but the fourth is often vegetative.

In many ways, the most fascinating change in the life of a tree is the onset of reproductive activity. Some species, notably pioneers, start flowering profusely after about 2–5 years, although the great majority remain purely vegetative for much longer than this. Nevertheless, specific precocious individuals are often noted in the literature as producing flowers before they are two years old, for example, *Anthocephalus chinensis, Dipterocarpus oblongifolius, Citrus* spp., *Funtumia africana, Hildegardia barteri, Monodora tenuifolia, Pinus caribaea* var. *hondurensis, Trema guineensis*, and many mangrove species. If such precocity is primarily genetic in origin, clones made from early-flowering individuals might provide an elegant research tool for miniaturising the study of flowering (Longman 1985). For commercial plantations, although early flowering may be an

Fig. 5.23 Six buds in the axil of one leaf of a tropical fruit tree in Sarawak. Note also the extra-floral nectary on the petiole, and the clearly marked site of the separation layer at the leaf base.

advantage in fruit trees it is probably undesirable in species that are grown for vegetative yields. There is some evidence that there has been unconscious negative selection in forestry, perhaps because seed is more conveniently collected from precocious and prolific individuals or stands. In teak, in West Africa, for example, entire young plantations may be seen flowering or fruiting copiously (Fig. 5.24), and occasional individuals may even do so in the nursery when only 3 months old (Hedegart 1976). Since the inflorescences are large, and terminate the main stem, this may be expected to cause poor form and reduced vegetative growth.

In most forest trees, however, there is a more extended **juvenile period** through which they pass before flowering commences. It is common for this interval to average between 10 and 30 years, although there is considerable variation between sites and amongst individuals (Ng 1966; Longman 1985). In general, it tends to be shorter in plantations than in natural forest, probably for a combination of genetic and environmental reasons; and it is also typically a briefer stage for pioneer species than for climax species. Thus Ashton (1969), for example, considers that it is

Fig. 5.24 Precocious onset of fruiting in a young plantation tree of *Tectona grandis* at Onigambari, S W Nigeria. Large terminal inflorescences have been formed at a height of 3–4 m in the plant on the right, limiting the length of straight bole that will be produced. Compare with the other teak (left) which is still vegetative, and the larger *Triplochiton scleroxylon* in the centre.

often about 60 years before the majority of the dipterocarps of the upper canopy of Malaysian forests start to flower.

In one group of plants the onset of flowering is of particular significance – monocarpic (hapaxanthic) species which flower once and then die. *Agave* is a well-known example, in which the life-cycle is about 10 years, while in *Tachigalia versicolor* it ends after many years (Foster 1977). Thus monocarpy is not confined to species such as *Corypha* palms, which exhibit Corner's model (with a single vegetative axis), but can occur in emergent trees with many shoot apices. Another striking feature of this group is that they often flower gregariously (see section 5.9); and since clones that have been dispersed to different regions still tend to flower and die at the same time, some long-term endogenous 'calendar' is implied (Simmonds 1980). Some but not all bamboos are monocarpic, with the life-cycles often rather clearly known, ranging from a few years in *Chusquea tonduzii* to 80 years in *Rhipidocladum pittieri* (Pohl 1974). Such very long juvenile periods are thought to have evolved as extreme instances of deprivation followed by saturation of seed predators (Janzen 1976*c*).

More experimental studies are needed on the various physiological factors which may be involved in controlling the length of time a tree produces only vegetative shoots, particularly because it is often desirable to be able to shorten this in crop species. However, it is known that coffee is a short-day plant in its floral initiation, whereas *Stylosanthes guianensis* is a long/short day plant (Ison and Humphries 1984). In *Bougainvillea* spp. and *Euphorbia pulcherrima*, the first flower production in these vegetatively propagated ornamentals is affected by an interaction between photoperiod and temperature.

Cuttings originating from young seedlings of *Triplochiton scleroxylon* have flowered on several occasions under warm, long-day conditions in glasshouses in Scotland, at an age of 2–6 years from germination (Fig. 5.25) (Leakey *et al*. 1981). Observations on the nut-bearing *Shorea penanga* in a nutritional trial in Sarawak suggested that profuse flowering and fruiting at 6 years occurred particularly in plots receiving all the fertilisers (Lees 1980). Immature mango shoots have also been induced to flower and fruit by treating them with potassium nitrate (Bondad and Apostol 1979).

The most generally applicable treatment for stimulating forest trees to flower may prove to be growing them as quickly as possible, which appears to hasten the processes involved in maturation (Hackett 1985; Longman 1986). However, various additional techniques may be needed once the trees have reached

Fig. 5.25 Induction of early flowering in a potted *Triplochiton scleroxylon* in a tropicalised glasshouse at the Institute of Terrestrial Ecology, Edinburgh. One flower is just opening (in the evening), and the sites of abscission of some of the flower buds can be seen (arrows). Following hand pollinations, ripe fruits, a new generation of seedlings, and early flowering in some F1 individuals have been obtained. In this species, of the order of 0.1 per cent of flower buds normally set fruit.

'ripeness-to-flower', to provide them with the 'opportunity-to-flower'. Methods which have had some effect in seedlings include removal of the terminal bud in mango, decapitation in coffee just after the first branches are formed, and defoliation (using urea sprays) in guava. Bark-ringing has also been successfully used in a number of tropical as well as temperate woody plants.

Some indication of possible hormonal effects is suggested by the induction of flowering in some 10 year old cocoa trees, that had never been seen to flower previously, by transplanting into their bark small clusters of flower buds from very floriferous trees (Naundorf 1954). Treatment with gibberellins stimulates cone initiation in young trees in the Cupressaceae, and the techniques may be applicable to other conifers (see section 7.4). Other plant growth regulators may also be involved; for example compounds which release ethylene in the tissues have stimulated the onset of flowering in mango (Chacko *et al.* 1976; Shuzeng and Zongwe 1981). The herb *Geophila renaris* remains vegetative as long as the soil moisture content is near its maximum value. If it falls below the permanent wilting percentage (see Fig. 3.12) flowering is stimulated, and it continues afterwards even if the soil again becomes saturated with water (Bronchart 1963).

5.9 Flowering

Flowers often attract the attention, of people as well as of potential pollinators. However, many open only for short periods, and may be hard to see high up in the euphotic zone. There is an extraordinary range of shapes, sizes and distribution of flowers in the tropical forest (Figs 5.26, 5.27 and 7.18), with diverse colours, markings and often scents. *Raffesia arnoldi* forms a single, foul-smelling flower about a metre in diameter; cocoa bears hundreds of small, cauliflorous blossoms; *Bombax buonopozense* unfolds red, waxy 'ponds' high up in the canopy, in which aquatic invertebrates and vertebrates can flourish; while a large dipterocarp tree might remain purely vegetative for many years, and then suddenly produce a million flowers that all open within a week or two. With diversity on this scale, it is obviously difficult to generalise, although fortunately there is now a growing literature on floral biology at the community and species levels.

Fig. 5.26 Axillary flower buds and opening flower of *Xylopia aromatica*.

The critical first stage in reproduction is the conversion of an apex from producing only vegetative primordia to forming an inflorescence, cone or single flower instead of or in addition to foliage leaves. The first detectable signs are often an alteration in the shape of the apical dome, and in the growth rates of leaf and bud primordia, as well as inhibition of the sub-apical meristems which produce and extend the internodes. Floral development may continue without a break from initiation until the opening of the flowers, as for example in *Hibiscus* and many herbaceous species. However, in many trees there is an interval during which the initials remain within a dormant flower bud, which can vary from a few weeks to many months, or even 2½ years in some *Eucalyptus* spp. It is therefore necessary to distinguish carefully between the timing of floral initiation, and that of flower opening, since the factors which influence these two processes may not be the same (Borchert 1983; Longman 1985).

More or less **continuous** flowering (after the juvenile period has passed) is found in some evergrowing secondary tree species such as *Dillenia suffruticosa, Harungana madagascariensis*, and *Trema orientalis*. The common mangrove *Rhizophora mangle* also flowers continuously, although in Southern Florida it shows

Fig. 5.27 *Dissotis entii* of the Melastomaceae, a characteristic family of the undergrowth in W African rain forests.

a peak in June/July (Tomlinson 1980). In the great majority of cases, however, the flowers emerge **periodically**, either regularly or irregularly, though there is usually sufficient overlap for the forest as a whole never to be without flowers. Individual trees may also show variation within the crown, with some parts in flower, other branches perhaps in fruit, and others again producing only leaves.

On the other hand, synchronised flowering can occur over many hectares of forest, as in *Pterocarpus indicus* in Malaysia, *Tabebuia serratifolia* in Surinam and the West African liane *Calycobolus heudelotii*. A few species that flower annually are so reliable in their timing that they have been used as a signal for the planting of a food crop – for example, *Erythrina orientalis* for yams in the New Hebrides, *Sandoricum koetjape* for rice in West Malaysia and *Trichilia heudelotii* for the second planting of maize in Ghana. Perhaps they have proved more useful than astronomically based calendars, since they co-ordinate the agriculture cycle with current environmental conditions rather than with fixed dates, which do not allow for year-to-year variation in temperature and precipitation (Nations and Nigh 1980).

The frequency with which flowering periods recur varies from about 3–4 months in *Ficus sumatrana*, for example, to 10–15 years in *Homalium grandiflorum* (Medway 1972; Holttum 1953). Guava can flower twice a year, while fairly regular annual flowering occurs in the fruit trees *Baccaurea parviflora* and *Xerospermum intermedium* (Yap 1982), and in a number of common forestry plantation species, including *Cedrela odorata, Gmelina arborea, Tectona grandis, Terminalia ivorensis* and *Pinus kesiya* (Longman 1985). Biennial tendencies appear to be characteristic of most species reaching the canopy in Surinam forests (Schulz 1960), and may also become marked in older mango trees.

Although dipterocarps may occasionally flower again after 1–2 years, the interval is more commonly about 3–8 years (Wycherley 1973). In these irregularly spaced 'mast' flowering and fruiting seasons, the majority of the dipterocarps in a whole region generally flower profusely and gregariously over a period of a few months, often accompanied by heavier than usual reproductive activity by trees in other families. In a detailed study of six related *Shorea* spp. which involved frequent assessments high up in the canopy, Chan (1977) found that each species occupied a specific 'time-niche', with the flowers open for about 15 days in the first to flower (*S. macroptera*) and for about 25 days in the last (*S. leprosula*). Peak flowering times of the 6 species did not overlap at all (Fig. 5.28). This remarkable characteristic, which has also been found in the New World tropics for 4 *Guarea* spp.

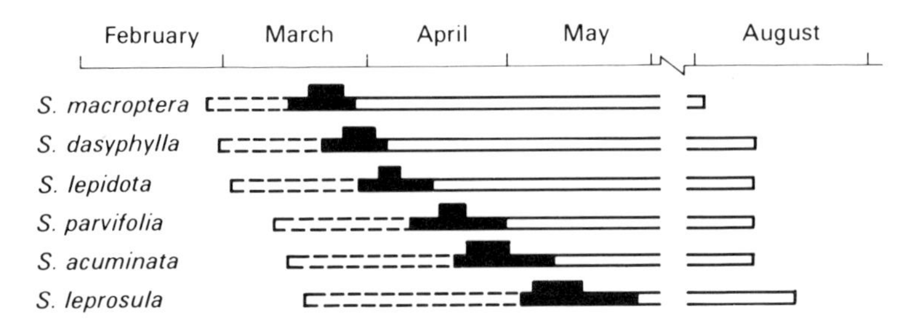

Fig. 5.28 Sequential flowering of 6 related *Shorea* spp. in the gregarious dipterocarp 'mast' year of 1976. **Dotted** zones = flower buds detectable with binoculars; **dark** zones = flowers open (plus peak flowering periods); **open** zones = fruit development. (From Longman 1985, after Chan 1977 and Chan and Appanah 1980.)

(Frankie *et al.* 1974) and in the Bignoniaceae (Gentry 1974), is thought to have evolved through competition for a limited number of pollinators. There is considerable speculation about the nature of the physiological trigger for irregularly spaced, but community-wide flower initiation. Perhaps the most likely hypothesis (which awaits testing) is that an increase in the diurnal fluctuation of temperature, associated with drier spells of weather with clearer skies, stimulates floral initiation at the same time in all the six *Shorea* spp. which then vary in their rates of subsequent flower development (Longman 1985).

Because some tree species show reproductive activity that is clearly or probably related with seasonal changes does not mean that all of them do so. The proportion of 'not clearly seasonal' species was 14 per cent in south Florida (Tomlinson 1980), 26 per cent in a detailed study at Barro Colorado Island, Panama, that also included herbaceous plants (Croat 1975), and about 33 per cent in Ghana (Hall and Swaine 1981). Some examples of aseasonal tree species are *Anisophyllea corneri, Anthonotha macrophylla, Blighia sapida, Genipa caruto, Santiria laevigata* and *Trophis racemosa*.

Considering the forest as a whole, however, there are usually peaks of flowering at certain times of year, even in rather uniform climates (Richards 1952; Baker *et al.* 1983). Comparisons between data from different regions can be made using a **seasonal flowering index**, which is the ratio between the number of species flowering in the 6 month period with the most flowering, and that with the least (Longman 1985). Values between 1.2

and 4.9 have been calculated for consecutive 6 month periods, the degree of seasonality depending both on the magnitude of the difference and the amount by which the number of flowering species exceeds zero. Aseasonality at the community level has been claimed for a forest in West Malaysia (Putz 1979), but it is interesting that if the data presented are recalculated, there is 25 per cent more flowering in the months March–May and September–November than in June–August and December–February. This difference is statistically significant, and gives a seasonal flowering index (on a 3 month + 3 month basis) of 1.25.

Not surprisingly, there tend to be closer links between reproductive phenology and particular times of year in forests experiencing definite climatic seasons (Janzen 1976*a*; Frankie *et al.* 1974; Longman 1985). Very often the flowering peaks occur during the dry season and the early part of the rains (Fig. 5.29). Deciduous species tend to flower in the dry season, expanding flower buds that do not also contain foliage leaves (e.g., *Ceiba, Hildegardia*), or mixed floral and leaf buds at the usual time of vegetative bud-break (see section 5.1; e.g., *Delonix, Peltophorum*). Peak flowering in evergreen species occurred during the transition to the wet season, the inflorescences or single flowers often emerging with the new leaves, or perhaps being produced cauliflorously.

There is more variation found between different forests when climate is relatively uniform. Most flowering occurred from March to July in a six year study of 56 individuals (42 species) in West Malaysia (Medway 1972), whereas June and July were not

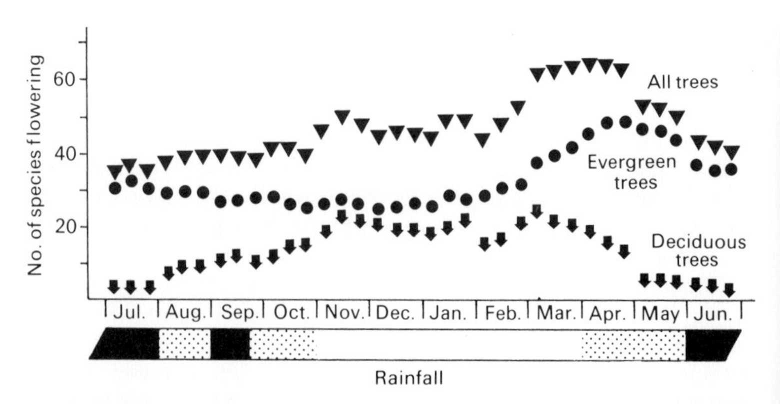

Fig. 5.29 Frequency of flowering (in 10-day periods) by 158 deciduous and evergreen trees in Ghana. (After Taylor 1960.)

good flowering months in a four-year study of 131 trees (62 species) only about 50 km away (Putz 1979). At Pasoh Forest Reserve, only about 150 km away, a 3½ year study of over 400 trees (19 fruit tree species) showed a clear predominance of flowering in the earlier part of the year (Yap 1982). A comprehensive study of the forests of Panama, involving three years fieldwork and the examination of more than 50 000 herbarium specimens, suggests that the different life-forms do not all flower at the same time (Croat 1975). For instance, trees, lianes and epiphytes showed peak flowering in the dry season, whereas shrubs and herbs growing in the undergrowth or in small gaps flowered especially in the rainy season.

The **timing of floral initiation** has been little investigated, although it is clearly an important topic that is relatively easy to study by dissection if suitable material can be collected regularly. This would be easier in species such as *Ateramnus lucidus, Bosqueia angolensis, Monodora tenuifolia* and members of the Cupressaceae, in which reproductive and vegetative shoots can be distinguished several months before flower opening occurs, because the floral buds are larger or the developing cones appear different. Besides clarifying the questions of seasonality of flowering discussed above, knowledge of when and where flower formation occurs in a tree can greatly assist physiological studies into the factors which promote floral initiation. For instance, if it is known that terminal or lateral reproductive organs start to form just as shoot elongation ceases (see Borchert 1983), experimental treatments can be applied with much greater temporal and spatial precision.

Coffee, tea, bougainvillea and poinsettia are all facultative or obligate short-day plants, and the suppression of flowering in *Hildegardia barteri* growing close to street lights has been interpreted in the same way (Hall *et al.* 1975). However, it should not be concluded that all tropical trees will be short-day plants or day-neutral, for a substantial minority of tropical plants have been reported to be long-day species (Mathon 1975), while temperature can modify effects of photoperiod. Drought stress has long been claimed to be a factor stimulating flower initiation (Schimper 1903), although the evidence is often lacking for tropical trees.

The most effective hormonal treatment for stimulating reproductive activity is undoubtedly the application of gibberellin (GA_3) to members of the Cupressaceae (Gutiérrez 1978; Longman *et al.* 1982). Large numbers of male and female cones can be stimulated, particularly if doses of about 1–100 mg are injected in alcohol into small holes drilled into the wood of suitable branches or main stems (see section 7.4). Other gibberellins ($GA_{4/7}$) appear to be promising tools for promoting cone initiation in *Pinus* spp., but in broadleaved trees it appears that gibberellins often inhibit floral initiation (Guardiola *et al.* 1982). Three different growth retardants promoted the formation of flowers in mango during an 'off' year, especially in combination with bark-ringing (Rath and Das 1979). Pronounced stimulation of flowering following a soil drench application of another inhibitor was found in the vine *Clerodendrum thomsonae* (Koranski *et al.* 1979).

Other treatments which have been found to modify flower initiation include bark-ringing, which is stimulatory in *Faurea speciosa, Lonchocarpus utilis, Pinus elliottii* var. *elliottii, Terminalia ivorensis*, mango and rubber (Longman 1985). It may also be possible to modify the sex of inflorescences of *Elaeis guineensis* towards male by removing the older leaves, although there is a lapse of two years before this is seen because of the long period of development after initiation. In *Carica papaya*, which is dioecious, there is some evidence that male trees can produce functional female flowers (though few seeds) if the stem has been damaged.

Progress in the physiological study of flower initiation is likely to come from a sustained effort to develop miniaturised systems in which flowering can be studied in depth (Longman 1985). In particular, the use of selected clones that can initiate flowers under controlled environments could transform our knowledge of the effects and interactions of external factors, and also elucidate the internal links and balances between the formation and growth of flowers, leaves, stem units, roots, etc. There is even the prospect nowadays of studying reproduction in aseptic cultures, in which even the hormonal balance of an organ can be carefully controlled.

The **opening of flowers** has been studied extensively in coffee, and also in trees and shrubs in a Costa Rican forest with a pronounced dry season (Opler *et al.* 1976). *Bernardia nicaraguensis* and *Croton reflexifolius* initiate and develop floral buds, and then become leafless during the dry season. Just 2–5 days after the first rains, they flower gregariously, whether this happens to be in March, April or May. Other woody species also appeared to be triggered by rainfall to open their flowers, taking longer to respond, but tending to occur each year in the same sequence. The shrub *Hybanthus prunifolius* similarly flowers synchronously in Barro Colorado Island, Panama, a few days after the first rainfall of 12 mm or more, provided that the preceding dry period has been sufficiently intense and prolonged (Augspurger 1982). Watering a plant after dry weather will stimulate it to flower, but watering it during the dry season prevents it from flowering.

A rapid drop in temperature normally accompanies a sudden tropical thunderstorm, and in coffee either a temperature-drop of at least 3 °C in 45 minutes, or the relieving of water stress by the rain, can trigger flower opening after a dry spell. In the epiphytic orchid *Dendrobium crumenatum*, the change in temperature appears to be the critical factor (Coster 1926); whereas with dry bulbs of the 'rain-flowers' *Pancratium* and *Zephyranthes*, which flower 3 days after heavy rain, a direct effect of moisture is indicated by their response to either warm or cool water (Holdsworth 1961). The environmental factors and changes in endogenous promoters and inhibitors that are associated with coffee flower opening are discussed by Browning (1977).

A surprisingly large proportion of flower buds may abort and be abscised before opening, particularly when there are large numbers present, as in *Shorea* spp. (Chan 1977) and *Triplochiton scleroxylon* (Leakey *et al* 1981). Sometimes the flowers on a plant all open within a week or so, as in the gregarious examples already quoted. In other species, such as *Jacaratia dolichaula* and *Pentagonia macrophylla*, a few flower buds may continue to open for as long as 3–4 months (Bullock and Bawa 1981; Augspurger 1983). Individual flowers may last for days, as in mangroves (Primack and Tomlinson 1980), or only for hours, as in many other tree species, and they frequently have very precise opening times that are clearly related to the behaviour of the appropriate pollinator. Thus, for example, anthesis starts around 1600 h and is completed by 2000 h in *Durio zibethinus*, while pollination (by the bat *Eonycteris spelaea* and a noctuid moth) occurs mainly between 2000 and 0100 h (Soepadmo and Eow 1976). The flowers of *Ceiba pentandra* open as dusk falls, and are pollinated by bats that take the nectar and distribute pollen on their fur (Baker and Harris 1959).

Many other physiological and structural aspects of flowers and pollination are covered by Tomlinson (1980), Bawa (1982), Tanner (1982) and Baker *et al*. (1983), amongst others. An important point is that around 30–40 per cent of tropical trees have been found to produce separate male and female flowers, the majority of these being dioecious, with separate male and female trees. Various intermediate categories also occur quite frequently, serving as a reminder of the great variability in reproductive biology which exists in the tropical forest.

5.10 Fruit development

Except in parthenocarpic plants such as banana, plantain and seedless varieties of *Citrus*, pollination must take place if fruit-set is to occur. Unpollinated flowers are usually abscised, and it is common for this to occur to many pollinated flowers as well. High proportions of flower abscission have been found for instance in *Aucoumea klaineana* (Grison 1978) and *Shorea* spp. (Chan 1977). Indeed, there can be several thousand flowers on a single inflorescence of *Parkia clappertoniana* or mango, and yet only 1–5 of them normally set fruit.

One reason for this may be that flowers which are fertilised first often inhibit fruit-set in those that are pollinated later (Browning 1985). On the other hand, in species such as epiphytic orchids that produce relatively few flowers, it may be noticeable that the individual flowers appear to set fruit independently of each other. External factors can also affect the setting of fruits: for example, water stress and insect attack are both likely to increase flower abscission (Addicott 1982). Either high or low temperatures partially or completely inhibited fruit-set in coffee, the optima being 23 °C day/17 °C night, or 26 °C day/20 °C night (Went 1957).

The subsequent enlargement and development of fruits takes place at rates that are characteristic of the species, modified both by the prevailing environmental conditions and by mutual competition between fruits and with other 'sinks'. By no means every fruit that is set reaches maturity, partly because fruitlet abscission continues to 'thin out' the crop remaining on the tree, even if there is no abnormal stress. Insect predators or mammalian herbivores may damage seeds or eat fruits before they are ripe, while diseases such as cherelle wilt in cocoa may cause a high proportion of the young fruits to shrivel on the tree.

The final stages in fruit enlargement are often relatively slow, but then ripening (senescence) can occur quite rapidly. During the latter process, a sudden rise in respiratory rates can frequently be observed in fleshy fruits, associated with the metabolic changes leading to softening and the development of the characteristic colour and flavour. The onset of ripening may be stimulated by an increase in ethylene levels within the fruit, associated with changes in auxins. In avocado for instance it has been shown that low levels of applied indoleacetic acid (1–10 micromolar) suppressed ethylene production and delayed ripening, whereas 100–1 000 micromolar applications stimulated ethylene production, respiration rates and ripening (Tingwa and Young 1975). Ripening can also be stimulated by picking the green, fully grown fruit.

Fig. 5.30 Large, heavy fruits of the cauliflorous tree *Omphalocarpum ahia*, which easily fall through the canopy to the forest floor. Note the separation zone on the nearest fruit.

Shortly afterwards, most fleshy fruits are abscised (Fig. 5.30), unless they have already been taken by primates, birds, bats, etc. On the ground, they are subject to consumption, dispersal and decay through the activities of many other forest organisms. Once again, only a relatively small proportion of the total survive, and sometimes the entire fruit crop may fail to produce a single viable seed. The same can occur if seed insects or fungi are widespread, as frequently occurs in *Triplochiton scleroxylon*, where the fruit is non-fleshy and single-seeded. Where such fruits contain many seeds, the ripening process involves drying, and abscission layers in the fruit wall generally allow the seeds to be shed while the fruit is still attached to the tree. In conifers, the ripe female cones open their scales as they dry. An example showing frequency of dispersal by different agencies is given in Table 4.6.

The period from anthesis to fruit ripening has been studied in West Malaysia from the literature and observations over a 2 year period (Ng and Loh 1975). The average interval was 4.3 months for 86 indigenous and 7 exotic species, with *Pterocymbium javanicum* showing the shortest period (3 weeks). *Firmiana malayana* and *Dillenia suffruticosa* (4–5 weeks) also ripened their fruits rapidly. Those taking 9 months or more to ripen included *Diospyros maingayi, Palaquium hispidum* and *Vatica ridleyana*. The New World species *Bertholletia excelsa, Enterolobium cyclocarpum* and *Pithecellobium saman* have also been reported to ripen slowly (see also Foster 1982*a*). In West Africa, *Lovoa klaineana* has a short fruiting season, and *Mimusops heckelii* a long one. *Bosqueia angolensis* and *Celtis* spp. may fruit twice a year, while *Trema guineensis* fruits continuously (Taylor 1960). Although there may be a good deal of within-species variation, data of this kind can help in planning fruit collections, based on the time since flowering, which is often relatively easy to detect. However, not every flowering event is followed by a fruit crop.

There are fruits growing and ripening all the year round in tropical forests, which is important for the survival of frugivorous bats, in particular, since they have to maintain their body temperature by regular feeding. However, there are usually **fruiting seasons** when a higher proportion of the species and individuals is in fruit. In Ghana, for instance, this is clearly during the dry season (Fig. 5.31), with a fairly steady build-up to March. Most seeds will therefore be dispersed before the start of the rains. In a forest at 850 m altitude in eastern Zaire, the 200 species showed least fruiting (8 per cent) in May, at the end of the 9 month wet season, and most fruiting (46 per cent) in February (Dieterlen 1978).

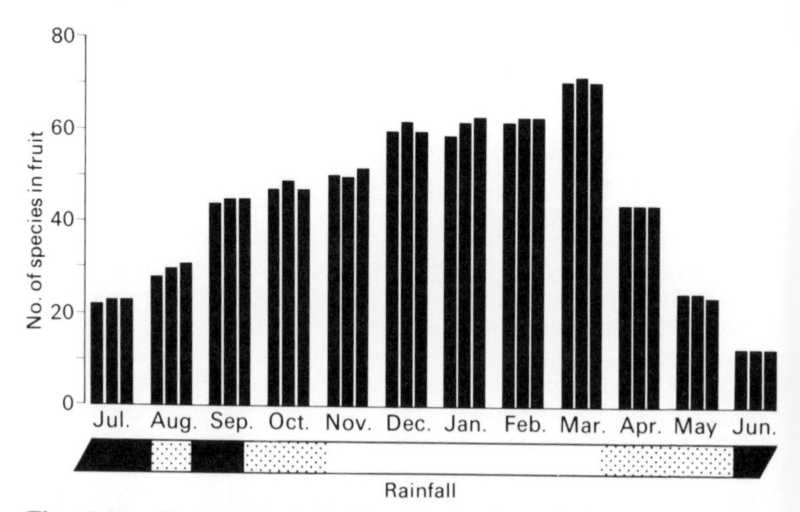

Fig. 5.31 Fluctuation in fruiting levels through the year in 158 forest trees in Ghana. (After Taylor 1960.)

Two dispersal peaks in the year were found by Foster (1982*a*) for the canopy trees at Barro Colorado Island, Panama, one in March/June (at the end of the dry season and start of the wet season) and one in September/October (during the rainy season). Lianes ripened fruit once a year, primarily between March and May, while understorey trees and shrubs produced most ripe fruit in November/December. In discussing the significance of these seasonal peaks, Foster recognises shorter and longer fruit development periods associated with species that flower at different times, such that they may disperse their seeds together. Also noted is the effect of weather variation, greatly affecting the animals normally depending on this food supply (Foster 1982*b*).

In West Malaysia, ripe fruits were most frequently seen in September and October, and were least common from January to May (Medway 1972). Of the 6 related *Shorea* spp. that flowered successively (see Fig. 5.28) the slowest rate of fruit development was found in the first to flower, and *vice versa*, with the result that all species bore ripening fruit at about the same time. Besides being a remarkable example of precise physiological control over the timing of reproduction, this result suggests that there may well be selective advantage in dispersing seeds at a particular time of year, even where the climate is relatively uniform.

However, by no means every genus distributes its fruiting periods in the same fashion – as with flowering there are instances of long-sustained fruit ripening, and of regular or irregular dispersal at intervals from a few months to many years. In West Africa, for example, fruits of *Terminalia ivorensis* have been collected from different sites in most months of the year, but on the other hand the fruiting of *Claoxylon hexandrum* is so regular that it is used to mark the date of a festival (Irvine 1961).

5.11 Seed germination and dormancy

A number of different types of seed germination can be recognised, based both on structure and function. In a study of more than 200 West Malaysian tree species, representing about 8 per cent of the indigenous tree flora, Ng (1978) found that the hypocotyl elongated in 72 per cent, carrying the cotyledons well above the ground surface. Most of these were the normal **epigeal** type, with cotyledons exposed and green, as in *Shorea leprosula, Koompassia malaccensis* and *Podocarpus neriifolius*, and many species that colonise clearings and margins. Two modifications were noted: in *Diospyros* spp. the cotyledons are abscised early, before turning green; and a **durian** category is recognised, for example, in *Durio zibethinus, Dipterocarpus oblongifolius* and *Strombosia javanica*, in which the cotyledons remain within the endosperm and are abscised later. They also stay inside the seed coat in the normal **hypogeal** type, in which the hypocotyl does not elongate significantly, as for instance in *Anisophyllea, Garcinia* and *Lithocarpus*. Here again, a modification is reported, with trees such as *Eugenia grandis, Lansium domesticum* and *Pithecellobium* spp. showing **semi-hypogeal** germination, with the cotyledons exposed at ground level.

Germination can also be classified in other ways (see Table 5.4). In a few species, the embryo has already developed into a viviparous seedling before becoming detached from the parent plant: *Dryobalanops aromatica* and several mangroves are examples of such **prior** germination (Fig. 5.32). A much larger number exhibit **prompt** germination, in which the radicle emerges within a few days or weeks of dispersal. This is characteristic of dipterocarps, for example, and many other tree species, particularly those having large seeds that contain considerable food

Fig. 5.32 Viviparous seedling of *Bruguiera gymnorrhiza*, still attached to the tree, with a massive radicle pointing downwards. Mangrove swamp near Bogamoyo, N W of Dar-es-Salaam, Tanzania.

reserves. Another class of prompt germinators are those species whose seeds germinate more rapidly and to a higher final percentage after they have passed through the gut of a herbivore. Such stimulation was found for instance in *Azadirchta indica*, *Nauclea latifolia*, and especially strongly in *Securinega virosa* seeds germinated from baboon dung in a dry forest/savanna woodland area in Ghana, compared with seeds taken from fruits that had not been ingested (Lieberman *et al.* 1979).

The simplest reason for **delayed** germination is that the seeds were formed in fruits that dried on ripening, and since being shed have not yet come in contact with moist soil or rain. All the other categories in Table 5.4 involve some type of dormancy, such that most or all of the seeds fail to germinate (although not dead) in moist conditions at normal ambient temperatures. This relatively simple idea, of seeds being 'dormant', is frequently confused in the literature with their being 'viable', or 'not germinating', concepts which are related but not synonymous.

The testa, and sometimes the indehiscent pericarp, may be impermeable to water until weakened by the combined effects of repeated cycles of heating and cooling, mechanical abrasion by mouth-parts or soil particles, chemical attack by micro-organisms, etc. Once water can penetrate, such seeds become imbibed and usually germinate at once, but the key ecological significance is often that the germination of a cohort of seeds is spread out over a period of months, years, or even decades, presumably

Table 5.4 Seed germination and dormancy

Timing	Class	Requirement	Examples
PRIOR	**viviparous**	none	*Ceriops; Dryobalanops aromatica; Magonia pubescens; Pithecellobium racemosum; Pouteria ramiflora; Rhizophora*
PROMPT	**direct**	none	cocoa; many dipterocarps; *Montezuma speciosissima; Mora excelsa*
	indirect	passage through gut may remove some dormancy	*Azadirachta indica; Nauclea latifolia; Securinega virosa*

Table 5.4 Continued

Timing	Class	Requirement	Examples
DELAYED	**dry**	water	
	dormant (dry seeds)	damage to testa/pericarp, and entry of water ('hard' seeds)	*Delonix regia; Enterolobium cyclocarpum; Hymenaea courbaril; Leucaena leucocephala; Ochroma lagopus; Parkia javanica*
		after-ripening at relatively high temperatures	*Elaeis guineensis;* rice
	dormant (imbibed seeds)	specific amounts and wavelengths of light	*Cecropia; Chlorophora excelsa; Macaranga; Musanga; Piper; Trema; Harungana madagascariensis*
		specific temperature regime	*Didymopanax; Ochroma lagopus; Heliocarpus donnell-smithii; Phytolacca icosandra*
		leaching of water-soluble inhibitors	tomato
		entry of oxygen	
		altered hormone balance	
		time for embryo to develop	

increasing the chances of some of the progeny surviving. 'Hard' seeds are common in the Leguminosae, although they are not formed for instance by *Koompassia malaccensis*.

Scarification (for instance by shaking with sharp sand or grit) is often used to pre-treat such seeds before sowing, and this increased germination of *Ochroma lagopus* from 3.5 per cent to 84 per cent (Vázquez-Yanes 1974). Surprisingly, effective techniques also included placing the balsa seeds in boiling water for up to a quarter of an hour, or subjecting them to dry heat between about 60 °C and 120 °C. It appears that this type of breaking of dormancy could be one of the reasons why balsa seedlings frequently spring up after the passage of 'light' fires.

Other seeds are freely able to take up water, but need to remain for a time in a dry condition for after-ripening to take place. Oil-palm, and freshly harvested cereals, including rice, generally do not germinate readily, but will do so after being kept indoors for a period. In the common annual grass *Dactyloctenium aegyptium*, dry storage at 20 °C for 10 weeks partially relieved this type of dormancy, while 40 °C was much more effective (Longman 1969).

Vázquez-Yanes and his colleagues have elegantly demonstrated and investigated light-sensitivity in colonising tree species in Mexico (see Vázquez-Yanes and Orozco Segovia 1984). The seeds of *Cecropia obtusifolia* and *Piper auritum* can remain dormant for more than a year if the imbibed seeds are kept in the dark. They respond to light by germinating, but unlike some of the classical responses of herbaceous species a substantial period of illumination is required. Nevertheless, the usual red/far red

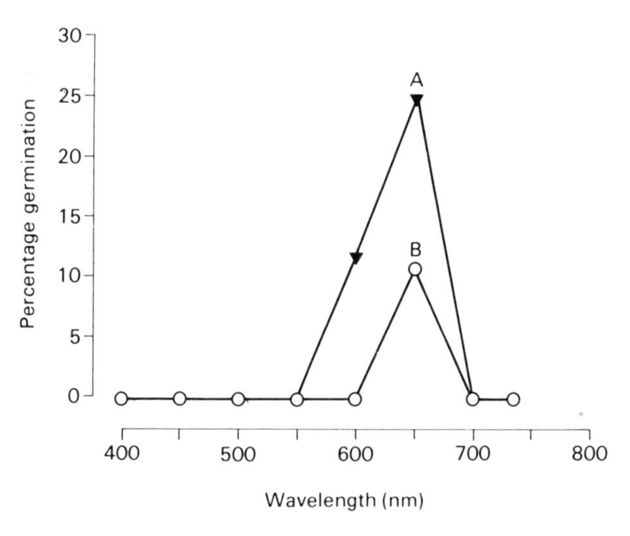

Fig. 5.33 Wavelength dependence of the breaking of seed dormancy by light, in (A) *Cecropia obtusifolia*; and (B) *Piper auritum*. Seeds were kept in the dark, except for 6 h irradiation at different wavelengths, and the final germination percentages were recorded after 1 month. (From Vázquez-Yanes and Smith 1982.)

system involving the pigment phytochrome is clearly operating, for red light is the effective wavelength (Fig. 5.33), and its stimulatory influence can readily be counteracted by following it with far-red irradiation (Vázquez-Yanes and Smith 1982). Even if red light was given for several hours, the germination percentage of *C. obtusifolia* only reached about 25 per cent, but values of around 90 per cent were achieved when seeds of either species were given 2 h of red light on each of 4 successive days. As discussed in sections 3.1 and 6.4, low red/far red ratios probably act to keep such seeds dormant within the forest (even in sun-fleck light) or in small gaps, but the higher ratios found away from the margins of larger clearings permit germination (Fig. 3.4).

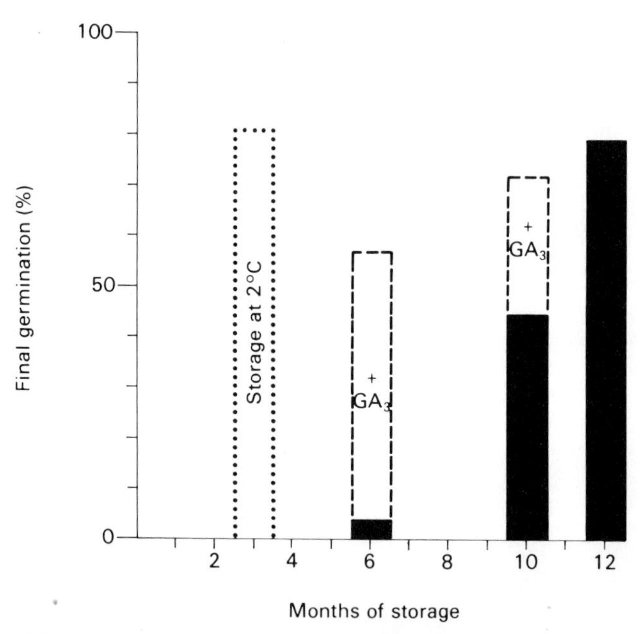

Fig. 5.34 Effects of storage time and temperature on the germination of *Trema guineensis* seeds, with and without gibberellin. **Black** columns = storage at 22 °C; **dotted** columns = storage at 2 °C; **dashed** column = promotion when seeds were germinated on agar plates containing 500 ppm gibberellic acid (GA_3). All germination percentages after 6 weeks in 12 h white light/12 h darkness at 22/28 °C. There was no germination at all in any of the corresponding 24 h dark series. (After Vázquez-Yanes 1977.)

The same group have demonstrated that a requirement for diurnal temperature fluctuation may be another mechanism involved in the breaking of seed dormancy of coloniser species in large gaps. For instance, imbibed seeds of *Heliocarpus donnell-smithii* mostly remained dormant when kept under constant temperatures in the laboratory or buried in soil within the forest at about 25 °C. High germination percentages were obtained under fluctuating laboratory temperatures, especially with a 6 h period at 32–39 °C, and 18 h at levels 10 °C cooler. Field germination was also enhanced near the middle of a sizeable gap, where the surface soil temperatures showed substantial diurnal fluctuations (Vázquez-Yanes and Orozco- Segovia 1984; see also section 3.2).

Further research can be expected to show whether other types of seed dormancy operate in tropical forest trees (see Table 5.4). Combinations of more than one type or class tend to occur in the same seeds, and it is not unusual for dormancies to change during storage. Nor is it always easy to correlate the results obtained in the laboratory with conditions in the forest. For example, *Trema guineensis* seeds stored at 22 °C would not germinate initially, but did so to an increasing extent the longer they were stored, at least up to one year (Fig. 5.34). The dormancy still present at 6 and 10 months could be substantially removed by placing the seeds on a medium containing gibberellin, suggesting that it might possibly have been mediated *via* a balance between germination inhibitors and promoters. Surprisingly, storage at 2 °C markedly promoted germination, rather as winter chilling breaks the dormancy of tree seeds in north temperate zone forests. Additionally, there was an obligate requirement for light, with no germination whatsoever being obtained in the dark series.

The germination temperature itself is also important, since all plants are thought to have minimum, optimum and maximum temperatures for this process, and germination rates are also strongly influenced by temperature. For instance, germination of coffee took three months at 17 °C, but only three weeks at 30 °C (Went 1957). Germination tests are generally carried out in convenient, standard conditions, for example at 25 °C or 30 °C, but as already mentioned these may not be appropriate for all species.

The same applies to the question of suitable storage conditions: sometimes the dormant state may remain unchanged for considerable periods, as for instance in seeds requiring after-ripening which may retain their dormancy if stored in a deep-freeze. More commonly, there is a continual modification of the degree of dormancy with the passage of time (Fig. 5.34). Storage

Fig. 5.35 View looking northwards from the side of a steep rocky hill near Ondo, S Nigeria, showing (foreground) leafless seedlings of *Hildegardia barteri* growing in a small crack in the granite. Farm land and secondary forest with oil palms in the background; note the haze typical of a 'Harmattan' period (December 1970).

at 50 per cent and 70 per cent relative humidity at 22 °C led to the development of a secondary type of dormancy in *Hildegardia barteri*, which was not the case when the fruits were stored at 90 per cent relative humidity (Enti 1968). This effect could be largely removed by placing the drier seeds into a moister atmosphere, and it apparently involves changes in the permeability of the structures surrounding the seed itself. It may be that this mechanism operates under natural conditions on the steep, rocky slopes colonised by this species (see Fig. 5.35). During the dry season, the majority of seeds may perhaps remain unresponsive to occasional rain or dew, but in the moister conditions of the rainy season they may lose their dormancy, and germinate with a greater chance of survival.

Storage temperatures and humidities are therefore the main considerations, together with avoidance of pests and diseases, when seeds are to be kept for use at a later date, particularly if long-term storage is planned for conserving gene resources. An unusually extended longevity of nearly half a century has been reported for *Ochroma lagopus* seeds from herbarium sheets (Vázquez-Yanes and Orozco Segovia 1984). However, in 'recalcitrant' species, which include many of the prompt germinator class of Table 5.4, viability can be quickly lost if the seeds are dried, and they either germinate or go mouldy if kept at high humidity. Nevertheless, it now seems likely that more thorough research may indicate ways in which some of these species may be stored (King and Roberts 1979).

For example, viability can be preserved for a month or more (instead of just a few days) if dipterocarp fruits are disinfected and stored at about 95 per cent relative humidity in closed containers (Soepadmo and Eow 1976). Temperatures must be above 15 °C for *Shorea ovalis*, and above 4 °C for several other species, of which *S. talura* was the most tolerant (Sasaki 1980). For oil palm seeds, formerly considered to be recalcitrant, the excised embryos can be thoroughly dried and kept at −196 °C, where it appears that viability may be preserved indefinitely (Grout *et al.* 1983). *Agathis hunsteinii* seeds keep better at a moisture content above 32 per cent, and die if dried to about 14 per cent, but *A. cunninghamii* appears to be relatively easy to store (Tompsett 1982). In *Triplochiton scleroxylon*, the speed of drying is an important consideration: when fruits were dried to 8 per cent moisture content at a rate not exceeding 1 per cent of the original weight lost per hour, they could be stored at −18 °C for at least 18 months (Howland and Bowen 1977).

6 Dynamic forest ecosystems

In Chapter 3, the environment found in tropical forests was described, and its influence on them considered both as regards average conditions and exceptional events. In Chapter 4, the complex and varied structure of these species-rich communities was analysed, and examples were given of the characteristics of the trees and other organisms which live in them. In Chapter 5, the physiological responses of individual woody species to environmental and other factors were discussed in terms of the overall growth and development of their major organs, with the emphasis on experimental investigations in controlled conditions. These three strands now need to be brought together, and linked with the question of soil conditions dealt with in section 2.3, in order to consider the tropical forest as a single, integrated and dynamic system that needs to be viewed as a whole (Fig. 6.1).

At the level of the whole ecosystem, one of the most useful concepts is that of the cycling of components. Besides immediately conveying the idea that the relationships within it are not static, a second important implication is of finite quantities of components being re-utilised, rather than of endless supplies being available. Moreover, the inputs, losses, consumption, storage, etc. may be estimated for each element of the system, providing a visual and quantitative flow-chart or budget. This can be done, for instance, with the flow of energy within the system, or that of water, of an individual mineral element, or even of genes.

Thus section 6.1 deals with **energy** balances, the various biological definitions of growth, primary and secondary production, biomass and yield. Section 6.2 covers litterfall, decomposition of biomass, and the recycling of **materials** that is carried on especially by the many small organisms in the soil. Section 6.3 gives examples of the sometimes remarkable ways in which the spread of genetic **information** from one population can be affected by interactions with other living components of the forest community, as well as with the abiotic part of its habitat. Putting together the cycling of energy, materials and information, section 6.4 discusses the natural damage and repair systems that lie

Fig. 6.1 View of the extensive area of secondary tropical forest which now covers the former Mayan town of Cobá, Yucatán, Mexico, seen from a high vantage point on an old temple.

behind the replenishment, succession and resilience of tropical forest ecosystems.

Synecological research of this kind can be greatly helped because modern electronic sensors, data-loggers, computers and visual display units can collect, store, process and present much larger and more complex sets of data than could be handled without their aid. Theoretical models dealing with part of the functioning of an ecosystem may also be constructed, concerning for example the cycling of a mineral nutrient, or the changes in species composition that might be expected in a gap. The usefulness of such models can be increased by examining how sensitive their predictions are to stepwise modifications of each of the basic assumptions, and how closely they tally with actual measurements. Great progress is likely in this field, provided that it is remembered that much simplification and extrapolation lie behind the smoothed curves, the calculated correlations and the prediction of principal factors.

Thus it is important to recognise the limitations, qualifications and uncertainties which surround every attempt to gain a synoptic view. It is perhaps salutary to remind oneself from time to time of the innumerable links that must exist between all the plants, animals and micro-organisms in the forest under study, as well as those between the living things, aerial environment and soil. In a very real sense, the science of ecology, which seeks an understanding of overall integration, is (like the observer standing on the floor of the tropical forest) fundamentally incapable of coping with the scale. In real life, the scientist has to content himself with small, unconnected fragments (while remembering that they actually form a whole); or with dimly perceived hierarchies of inter-relationships (while recalling that much diversity may exist between and even within the individuals of a single species).

In this spirit, the attempt in section 6.5 to capture such an elusive subject in the form of relatively hard-and-fast classifications of forest types may be seen as a useful tool rather than as a limitation on the imagination. Taken in this light, it may perhaps also help towards the good communication between scientists and the users of tropical forest land that is stressed in Chapter 7.

6.1 Energy, productivity and biomass

The simple statement that virtually all the energy in an ecosystem originates from the sun has a number of important implications. Firstly, each of its living organisms is therefore completely dependent on photosynthesis by the green parts of plants. Secondly, they all similarly rely on the temperature regimes that result from the net solar radiation balance at the Earth's surface. Thirdly, and rather unexpectedly, when there are sizeable inputs by man and his machinery to a managed ecosystem, it could be misleading to ignore the indirect energy subsidies contributing to subsequent harvests – for example, imported food and fossil fuels (both also derived from solar energy).

Some of the basic aspects of radiation receipt by tropical forest canopies have been described in section 3.1. Around 300 kJ cm^{-2} yr^{-1} of energy reaches the forest, but only about 1 per cent or less of the visible light received is utilised in photosynthesis. The maximum rates for this appear to lie between about 21 and 26 mg CO_2 dm^{-2} h^{-1} for pioneer species, 6–24 units for rain forest trees and 2–4 units for herbs and woody plants of the undergrowth (Bazzaz and Pickett 1980; Medina and Klinge 1983). Fig. 6.2 shows how measured rates of photosynthesis changed with increasing photon flux density in different groups of tropical forest plants. Clearly the pioneer example achieves much the highest rates in bright light, but it also appears to utilise dim

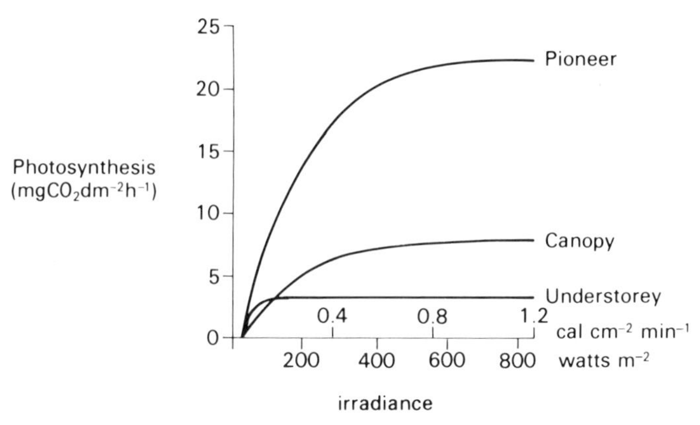

Fig. 6.2 Relationship between rate of photosynthesis and light intensity in examples of 3 groups of New World tropical species. (After Bazzaz and Pickett 1980.)

light more efficiently than canopy species. Understorey plants soon reached a maximum rate. The great majority of the plants in the tropical forest seem likely to be C_3 species, photosynthesizing via the normal Calvin cycle. The only examples so far reported of the more productive C_4 type (found in sugar-cane and utilising the additional highly efficient capture of CO_2 by the enzyme phosphoenolpyruvate carboxylase) are some woody arborescent *Euphorbia* spp. in low, open rain forests in Hawaii.

The same enzyme is also very active in plants with Crassulacean Acid Metabolism (CAM) allowing them to collect considerable amounts of CO_2 at night through open stomata, and store it in the large vacuoles in the form of malic acid. Stomata are often closed during the hottest part of the day, allowing water to be conserved, but photosynthesis can still proceed using CO_2 released from these temporary stores. A large number of American and north-east Australian tropical forest epiphytes show CAM metabolism, including bromeliads (Smith *et al.* 1985), cacti, many orchids and some ferns, suggesting that this is an important specialised adaptation to periodical drought stress. A recent estimate puts the number of CAM species at 10 000 to 15 000, which is about 5 per cent of the world's angiosperms.

Not all the photosynthesis of the forest is carried out in the laminas of leaves. Petioles and young stems are often green, while many older twigs have a zone of chlorenchyma immediately beneath a pale-coloured layer of cork. Stems and petioles are the main assimilating organs in species forming cladodes or phyllodes, such as *Acacia mangium*. Even roots may contain chloroplasts, and aerial roots can sometimes contribute substantially or even entirely to the plant's photosynthesis (Fig. 6.3). Greenish flowers and cones occasionally occur, while it is thought that many growing fruits can contribute to a greater or lesser extent to their own energy requirements. More than half the perennial species investigated had chlorophyllous embryos in a study of seeds developing during the dry season in a Costa Rican deciduous forest (Janzen 1982).

Fig. 6.3 *Microcoelia caespitosa*, an epiphytic orchid of the W African forest, in which thick aerial roots carry out most of the photosynthesis. The shoot system is greatly reduced, apart from the inflorescences.

Because of the infrequent occurrence of stomata close to such chloroplasts, the contribution of such organs is often dismissed as trivial. Nevertheless, it is possible that they could refix considerable amounts of respiratory carbon dioxide, despite having little gas exchange with the exterior atmosphere. Recycling of carbon within the tropical forest is also suggested by evidence that understorey herbs may utilise carbon dioxide from 'soil' respiration. Moreover, contrary to expectation, some of them appear to conserve assimilates through respiration rates as low as 0.2 mg CO_2 dm^{-2} h^{-1} (Löhr and Müller 1968).

Productivity is a concept that causes much confusion, since it is used in many different senses. At the ecosystem level, the basic biological definition regards all the photosynthesis occurring in the forest over a period of time as the **gross primary production** (GPP), often expressed in t ha^{-1} yr^{-1}. Some of the energy (and carbon) fixed in photosynthesis is released again during respiration by the green plants for their own growth and maintenance. If this respiration is deducted from the GPP, it gives the **net primary production** (NPP), one of the most valuable of ecosystem parameters. It represents the amount of energy available to all the heterotrophic members of the community. A slight confusion may be implied by the use of the word 'net': it does not mean that this is the quantity available to be harvested by man, for the animals and decomposers will have consumed some or all of it.

The NPP values that have been calculated for tropical forests (see also Fig. 7.1) are amongst the highest values for any natural ecosystem, and are surpassed only by a few very rich grasslands and subtropical plantations. Summarising a number of studies, Murphy (1975) found an average of 22 t ha^{-1} yr^{-1} (with a standard error of about two) for rain forests, higher than for other tropical ecosystems and 70 per cent greater than for temperate forests. More recent reviews confirm this and suggest that values can range from below 15 to over 30 (UNESCO 1978: p. 242; Medina and Klinge 1983). Strictly speaking, most NPP estimates include a small quantity of mineral elements incorporated into the products of photosynthesis – if the ash content is excluded then the data refer to **organic production**. If the roots are omitted, then the values are termed **above-ground production**. The weight of new tissue which is added to the 'standing crop' of animals in the ecosystem is called the **secondary production**. Though it is difficult to estimate, it is generally only a small fraction of the NPP. The standing crop is the **biomass**, or total amount of living organic matter present in animals at any one time.

Plant biomass figures for tropical forest ecosystems are also amongst the highest in the world, averaging about 415 t ha^{-1} according to a summary by Golley (1983*b*). Of this almost 90 per cent may be in the stem, with only just over 2 per cent in the leaves and 9 per cent in the roots. However, the root component rose to more than half in a mangrove forest, because of the allocation of assimilates to the still-roots and the relatively low and slender stems. The total biomass ranged from about 350 to 550 t ha^{-1}, with an exceptional record of 1 185 for riverine forest in Panama. Lower values than those quoted were found in swampy or dry conditions. Although the wood of different species varies widely in density the actual energy values of various parts of tropical trees were remarkably similar. Averages are usually between 15.5 and 18 kJ g^{-1} dry weight, not differing appreciably from one type of forest to another, or in comparison with temperate vegetation.

In a regenerating stand or a young plantation, NPP normally considerably exceeds decomposition plus consumption by herbivores, and so the plant biomass increases annually. Eventually, however, a secondary forest can be expected to reach a steady state, in which biomass remains roughly constant. In stable primary forests too, an equilibrium has usually been attained such that on average the same amount of energy is lost from the system (through the decay of leaves, fallen trees, other litter, dead animals and micro-organisms) as is made available within the green plants. In one sense nothing is being produced, while from the point of view of ecological energetics such tropical forests are, as already stated, amongst the most productive ecosystems in the world.

However, as C. F. Jordan (1981; 1983) has pointed out, the proportion of this high rate of dry matter production that is allocated to stem wood is relatively lower than in temperate zone forests. Table 6.1 shows that though a tropical forest had a much greater biomass in wood and leaves than its temperate broad-leaved partner, its respiration losses and maintenance costs for leaves were also higher (see also Monteith 1972). This suggests that there will not necessarily be a particularly high annual production of wood by tropical trees, in comparison with temperate forests.

There are very many other ways in which growth, biomass, productivity and yields are assessed (Table 6.2). Botanists often use growth analysis to examine the efficiency with which light is absorbed, carbon dioxide fixed and dry weight accumulated by

Table 6.1 Contrasts between biomass, respiration losses and maintenance costs between a temperate and a tropical forest. (From Medina and Klinge 1983)

Forest	Place	Above-ground biomass (t ha^{-1})		Respiration losses (t ha^{-1} yr^{-1})		Maintenance costs (t t^{-1} yr^{-1})	
		Wood	leaves	Wood	leaves	Wood	leaves
46 yr *Fagus* forest	Denmark	129	2.7	4.5	4.6	0.035	1.7
Lowland dipterocarp forest	Pasoh, Malaysia	414	7.6	18.8	29.1	0.045	3.8

Table 6.2 Some common measures of growth, biomass, productivity and yield

Purpose	Measure	Example	Units
Ecosystem study	Gross primary production (GPP)	Weight of all photosynthates produced in the forest per unit time	$t\ ha^{-1}\ yr^{-1}$
	Net primary production (NPP)	Portion of GPP not used by producers for their maintenance and growth	$t\ ha^{-1}\ yr^{-1}$
	Biomass	Total dry weight of all forest organisms at a particular time	$t\ ha^{-1}$
Growth analysis	Net assimilation rate	Net gain in weight by CO_2 fixation per unit area of leaf in a given time	$mg\ m^{-2}\ day^{-1}$
	Leaf area index	Leaf area per unit area of land surface	$m^2\ m^{-2}$
	Absolute growth rate	Dry weight produced by plant in unit time	$g\ day^{-1}$
	Relative growth rate	Dry weight produced per unit weight of plant in a given time	$g\ g^{-1}\ day^{-1}$
Agriculture	Crop yield	Fresh weight of crop harvested per unit area	$kg\ ha^{-1}$
	Harvest index	Yield as % of above-ground biomass	%
Forestry	Basal area increment	Gain in cross-sectional area of tree at breast-height (1.30 m) in unit time	$cm^2\ yr^{-1}$
	Current annual increment (CAI)	Annual increase in volume of above-ground wood per unit area	$m^3\ ha^{-1}\ yr^{-1}$
	Periodic annual increment (PAI)	Same, averaged over a 2–5 year period	$m^3\ ha^{-1}\ yr^{-1}$
	Mean annual increment (MAI)	Same, averaged over the whole life of the stand	$m^3\ ha^{-1}\ yr^{-1}$

green cells and organs, or in whole plants and communities. On the other hand, a farmer will be interested in the yield of a crop per hectare, and a plant breeder in the harvest index, which is the proportion of total dry weight that the old and new varieties allocate to the part or parts to be harvested. Growing a longer-term crop, foresters often wish to predict the volume of timber they may expect to be able to harvest by making successive assessments of diameter, height and number of stems per hectare. From these measurements, estimates of the increment in basal area can be derived, and also the stem volume (over- or under-bark), if the average taper of the boles is known from sample trees or previous experience. Additionally, they may wish to take into account the time when the mean annual volume increment peaks and begins to decline when deciding how long a rotation to use.

An instructive thought to bear in mind is that each of these and other types of calculation provides only one glimpse into the many facets of growth and development proceeding in a natural or managed ecosystem. It has been stressed in Chapter 5 that the production, expansion and duration of the various organs of a tree are influenced by many environmental factors, by its particular genetical make-up, and by internal control systems of various kinds. One cannot therefore really expect to be able to capture the dynamic whole by measuring a single parameter.

Another important point is that the communication barriers between scientists and managers are likely to be increased if the implications and limitations of the measurements they are employing are not appreciated. The opposite is actually needed – to encourage each to translate their experience and problems into the other's language and situation. Amongst the questions that might be relevant to a manager, for instance, are the following:

(a) What difference does it make if wood is being produced through short-term coppice regrowth, medium-term plantations or longer-term stands?
(b) How might the situation differ if yields consist partly of fodder, fruit or returns through agroforestry?
(c) Can terms such as 'harvestable productivity' and 'useful productivity' be used to translate NPP measurements into a form appropriate for each type of yield?
(d) In what way can questions about the continuation of yields into the future, and about the many other functions of forests, be incorporated into decisions being taken?

How to achieve useful interactions between managers and scientists is further explored in Chapter 7.

For the biologist, attempting to unravel the paradox of how mixed tropical forests achieve high NPP on distinctly 'unpromising' sites is one of the crucial tasks to be addressed in the next decades. Can more of the large throughput of energy be utilised by man without destroying the intricate systems by which it is cycled?

6.2 Decomposition and nutrient cycling

Sooner or later every forest organism dies – mayflies live only a matter of days, while some trees may survive for several hundred years. At their death the biomass of each organism is decomposed into simpler organic and inorganic substances, liberating part of the energy it contains. A variety of physical, chemical and biotic processes contribute to the piecemeal breakdown of the larger structures, much of the starch, cellulose, chitin and other carbohydrates, together with proteins, lipids, etc. being consumed, mineralised or otherwise converted by the living members of the forest ecosystem. As this happens, dead tissues, organs and organisms tend to fall down from the above-ground space, and accumulate on the surface of the soil, which thus becomes a dominant location for decomposition, rather like a natural 'forest cemetery'. Compared with other terrestrial ecosystems, however, a relatively larger proportion of the dead biomass decays while still within the canopy. This happens partly because the conditions favour rapid decomposition while trunks and branches that have died, for example, still have not fallen; but it is also a consequence of the presence of crown humus (Fig. 6.4) and epiphytic 'gardens', which result in about 10 per cent of the total being decomposed high above the forest floor (Edwards 1977; Golley 1983*a*).

As described in section 6.1, the rates of biomass production can be higher in tropical than in temperate forests, especially in the formation of foliage and twigs in the crowns, and also under favourable conditions in the production of wood. In humid lowland and montane forests, about 5 t ha^{-1} of dry organic matter is released for decomposition each year. This input is about five times greater than the average for temperate zone forests, and the rate of conversion of this fresh organic matter to humus is also high – about 30–50 per cent per year (Golley 1983*a*). Leaf litter in tropical forests may decompose in 3–4 months, compared with the two or more years needed in forests at higher latitudes. Meentemeyer (1978) found a linear relationship between leaf litter decay and evapotranspiration, which tends to

Fig. 6.4 Crown humus collecting in the rosette of leaves on a *Dracaena adamii*. Note the aerial roots.

confirm the important role of the higher temperatures and humidity of the moist tropics.

Decomposition actually consists of a chain of successive processes in which a number of forest organisms take part. As one example, leaves of a tree are chewed by a herbivorous mammal such as a sloth, the dead tissues are attacked by digestive enzymes and by intestinal bacteria and protozoa, the faeces are reconsumed by nematodes or beetles, the frass from these coprophagous invertebrates is again exploited by soil-living bacteria, protozoa and fungi, until much of the available energy and carbon has been released, and only humus, minerals and water remain. Innumerable other breakdown chains exist, as in other ecosystems, but it is of interest to ask whether there are any substantial peculiarities to be found within tropical forests.

According to several ecologists (see Golley 1983*a*), the decomposer systems of tropical forests show a relatively greater importance for the bacterial and fungal components. In a broad, comparative study of several forest ecosystems, Kitazawa (1971) found the lowest biomass and numbers of soil fauna in the rain forest, thus indirectly stressing the greater importance of microbial decomposers. The densities of bacteria found by Witkamp

(1970) in the submontane rain forest in Puerto Rico were about 30–40 million per cubic centimetre of soil. The actual performance of bacterial decomposers depends to a marked extent upon the energy and nutrients available in the litter. Even a minute imbalance – for example, between phosphorus and nitrogen – may cause anomalies in the processes of mineralisation, and thus in the availability of nutrients to the roots of the plants.

In acid and podzolic soils, such as the majority of oxisols, soil fungi may be dominant among the microbial decomposers (Fig. 6.5). Little is known about their specificity, but Went and Stark (1968) have shown that the hyphae of mycorrhizal fungi can transport essential ions directly from decomposing litter to roots. Even if this important 'short-circuit' is not always operating, the extensive mycelia of such fungi contribute profoundly to nutrient cycling in tropical forests, since most forest tree species form vesicular-arbuscular mycorrhizas, and some of the Dipterocarpaceae and Caesalpiniaceae, for example, form well-developed ectomycorrhizas (Fig. 6.10; Janos 1983). In one sense the green plants of the forest are not, strictly speaking, fully 'autotrophic' organisms, a thought also suggested by the root-grafts of *Aucoumea klaineana* described in section 5.7.

Fig. 6.5 Characteristic fruiting body of the Basidiomycete *Dictyophora phalloidea* in the litter of a rain forest.

Although soil arthropods as a whole may be somewhat less prominent in decomposition processes, termites are insects that are omnipresent in tropical forests, and they clearly play a very important role. It is primarily because of their activity that decomposition of wood does not lag far behind that of the leaves that are so rapidly disintegrated. The ability of termites to break down wood and other plant litter is in fact greatly enhanced by the presence of symbiotic protozoa in their digestive tract. Cellulose, hemicelluloses and lignin, the principal components of wood, are split by the joint digestive enzymes of termite and protozoan into less complex carbohydrates which can be utilised by other decomposers. Thus, unless impregnated with natural repellents or toxins, decaying logs from large boles and limbs disappear remarkably quickly from the floor of the tropical forest. The estimated time for complete decomposition of a large tree in Panama was less than 10 years (Lang and Knight 1979).

One of the most important topics in tropical forest ecology, nutrient recycling, follows naturally from a consideration of the decomposition of dead biomass. As already explained in section 3.4, only occasionally (except in hydromorphic sites) are the highly weathered profiles of tropical soils rich in nutrients. Thus the explanation for the frequent presence of tall trees, strikingly luxuriant forest communities and high primary production lies in the existence of very efficient mechanisms for capturing and retaining within these ecosystems both the limited resources of nutrients contained in the soil, and those entering from the atmosphere. Such efficient nutrient cycling implies rapid uptake, economical utilisation and conservation against loss from the ecosystem, and has been recognised as one of the most striking characteristics of mature tropical rain forests (Golley 1983*a*; Jordan 1985).

The movement of mineral elements within the forest ecosystem is governed primarily by its biotic components – for example, by the incidence of various species of plants, animals and micro-organisms – or by the presence of particular producers, consumers and reducers, using E. P. Odum's simplified model. Golley *et al* (1975) divided the organs of producers (green plants) into sections according to palatability, and consumers by their food specialisation, and produced a diagrammatic model of the flow of mineral elements in a tropical forest ecosystem (Fig. 6.6).

Litterfall studies have been undertaken in several different parts of the tropics, in order to make quantitative estimates of the important pathways of mineral nutrient cycling, throw light on the efficiency of utilisation and detect any seasonal fluctu-

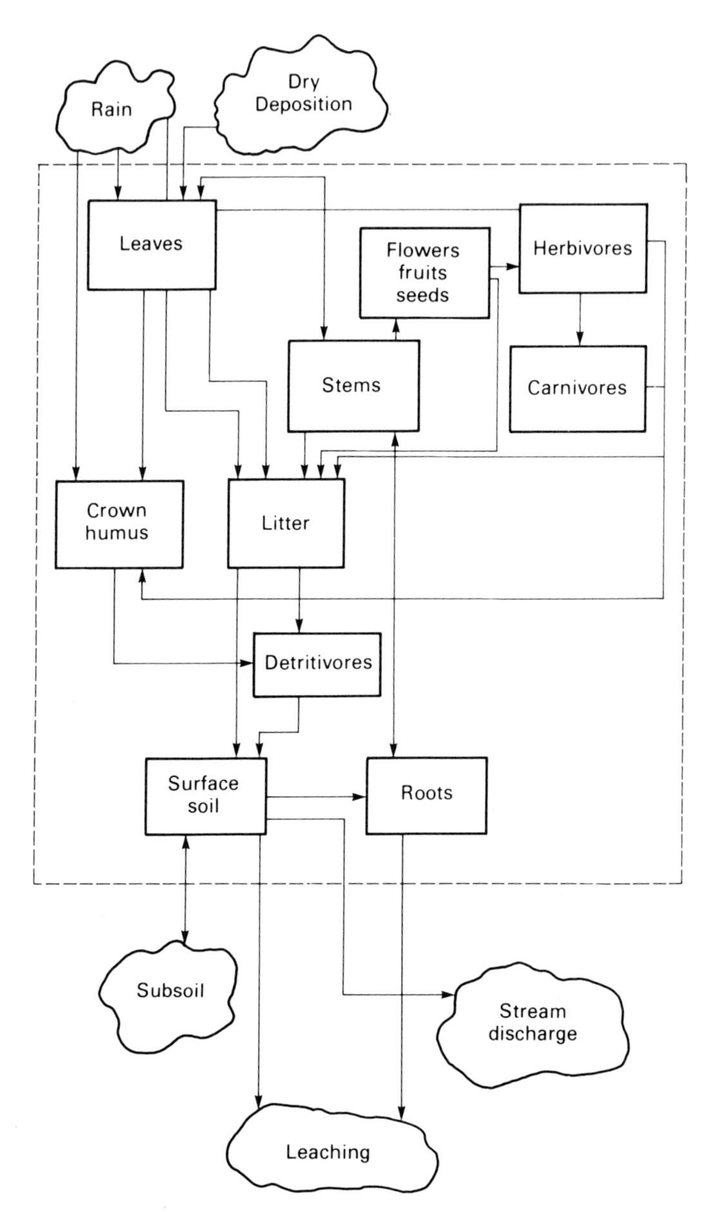

Fig. 6.6 Diagrammatic model of the cycling of materials in an ecosystem. (After Golley 1983b.)

ations (see section 5.5). Despite the fact that dead and dying biomass is distributed unequally through the three-dimensional structure of the forest, random litter traps situated on the ground may provide reasonably good samples of detached leaves, twigs, flowers and fruit, and of some small animal carcases and droppings. The amount of litterfall recorded in a period of time can then be compared with the amount of the **litter standing crop** lying on the floor in an adjacent plot. The ratio: litterfall/litter standing crop is called the **turnover coefficient**, and this provides an approximate and very simplified measure of the first stages of decomposition (Anderson and Swift 1983). Judging by the values of this coefficient that have been found, at least some tropical forests appear to show higher rates of decay of small litter than most temperate forests. However, many more measurements are needed before any general statement is possible (Proctor 1983). It also seems that the decomposition of 'timberfall' (the larger dead branches, limbs and trunks), the decay of irregularly spreading root systems and the 'rainwash' component of throughfall may combine to cause different patterns of mineral cycling than that which can be sampled by normal litter traps.

There are only a few complete investigations which provide estimates of the nutrient composition for all compartments of the biomass and throughfall. The first fairly complete picture was described for tropical forest in West Africa by Nye (1961). Table 6.3 shows the output of selected elements in litterfall and timberfall, suggesting a notably high nitrogen content in litter, which surpasses the values generally observed in temperate forests. As the nitrogen content of mature leaves in the crowns of tropical trees appears to be no greater than in temperate forests, the explanation may perhaps involve a more efficient migration of nitrogen during leaf senescence prior to leaf-fall in the temperate zone. It can also be seen from Table 6.3 that including the fallen dead wood may increase the estimates in the nutrient budget by

Table 6.3 The nutrient cycle under mature forest. (From Nye 1961)

Addition to soil surface from forest	Dry weight kg ha^{-1}	Nutrient elements kg ha^{-1} yr^{-1}				
		N	P	K	Ca	Mg
Litterfall	10 536	200	7.3	68	206	45
Timberfall	11 200	36	2.9	6	82	8
Rainwash		12	3.7	220	29	18
Total		248	13.9	294	317	71

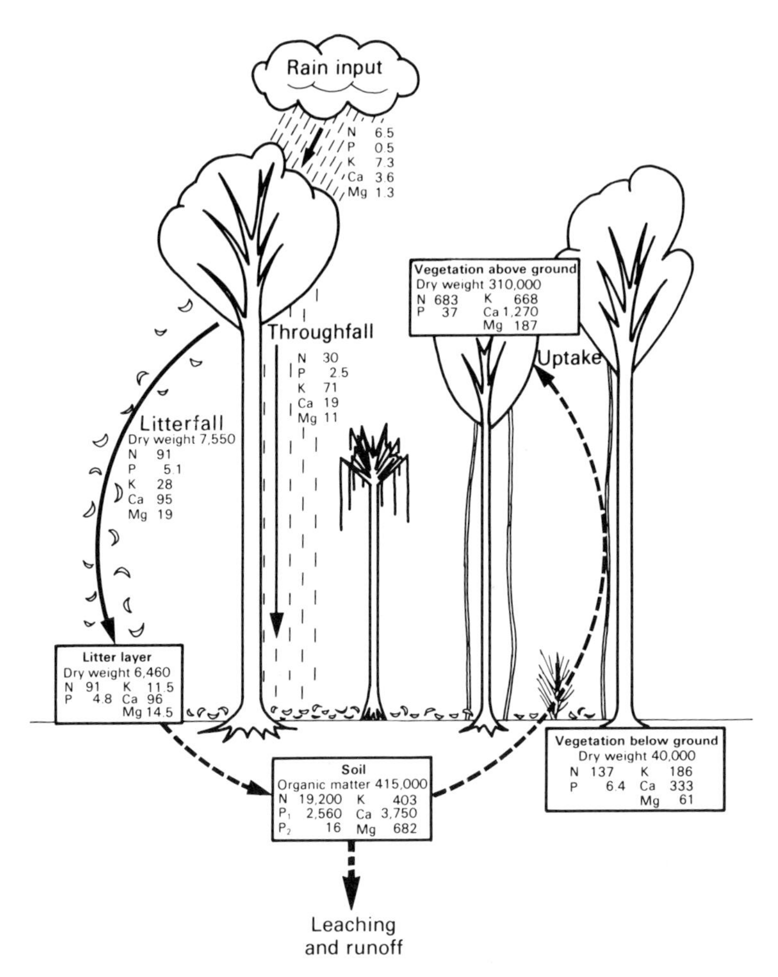

Fig. 6.7 Nutrient cycling in a montane rain forest in New Guinea. The data **in boxes** are the content of dry matter and nutrients in kg ha^{-1}; the figures by **black arrows** are the amounts of materials cycling in kg ha^{-1} yr^{-1}; **dashed arrows** are pathways lacking quantitative data. For the soil compartment of the ecosystem, total N, exchangeable cations, and total (P_1) and fluoride-soluble (P_2) phosphorus are shown. (From Edwards 1982.)

about one-third for calcium, one-quarter for phosphorus, one-sixth for nitrogen and one-eighth for magnesium. In comparison with the amounts falling in litter, the throughfall component, containing leaf-washings and some elements from the atmosphere (including dust, other dry deposition and nitrogen oxides produced by lightning), contained three times as much potassium, half as much phosphorus and magnesium, and only small quantities of calcium and nitrogen. The abundance of potassium in throughfall water is in agreement with the extreme mobility of this element which is known from temperate forests.

These pioneer ecological observations have been broadly confirmed in other parts of the tropics, but it seems that different types of forest may vary in their nitrogen cycling. In a detailed investigation in a montane rain forest in New Guinea, Edwards (1982) was able to draw up a fairly complete diagram of the major pathways of mineral transfer (Fig. 6.7). This showed very large amounts of nitrogen accumulating in the soil, and substantial quantities in the annual throughfall. It was concluded that nitrogen is probably fixed by micro-organisms in the phyllosphere, and that it becomes immobilised in the soil and rendered unavailable to plants. It is clear that much remains to be done before fuller pictures of nutrient cycling emerge, but it is already quite evident that the efficient systems which exist are often not maintained if the forest is removed (see section 7.2). Thus nutrient cycling on tropical forest land is much more a feature of the vegetation than of the soil which remains if the trees are removed.

6.3 Some biotic interactions

In this section, the emphasis is moved to the third type of cycling, involving the transfer of genetic information that occurs through the interplay of each living component of the tropical forest ecosystem on the others. As Charles Darwin wrote in 1859,

> *the structure of every organic being is related, in the most essential yet often hidden manner, to that of all other organic beings with which it comes into competition for food or residence, or from which it has to escape, or on which it preys.*

These relationships vary from the casual and transient to those in which the mutual bonds are long-lasting and essential for one or both organisms. They may be generally beneficial to both partners, or detrimental to one of them, as in many host/parasite, producer/grazer and herbivore/carnivore situations (see also Fig. 6.8).

Fig. 6.8 Pitchers of *Nepenthes* sp., growing on the floor of a rain forest at Semongoh Reserve, near Kuching, E Malaysia. The fluid in this insect trap contains digestive enzymes from the plant.

The last two are amongst the commonest of biotic interactions, forming the basis of the food chains in every ecosystem. They are usually not one-way relationships, however, for counterbalancing factors often exist, whereby a green plant produces extra tannins and lignins, which apparently act as 'digestibility-reducers' (Waterman 1983); or an animal species avoids capture by mimicry of the environment (see Figs 4.29 and 4.30). Sometimes a balance of positive and negative effects (mutualism) exists between the interacting organisms, as for example between frugivorous animals and the trees whose seeds they disperse. The few examples discussed below are mostly of this type, and are chosen because of the light they throw upon the diverse, unexpected and even bizarre links which exist between the plants and other living things in the forest.

A little-studied effect of one plant upon another is known as allelopathy, in which substances released into the local environment by one individual inhibit the survival or growth of others. Fig. 6.9 shows examples of laboratory experiments which suggest that the leaves of two of a group of four competing pioneer tree species may produce substances having the potential to reduce

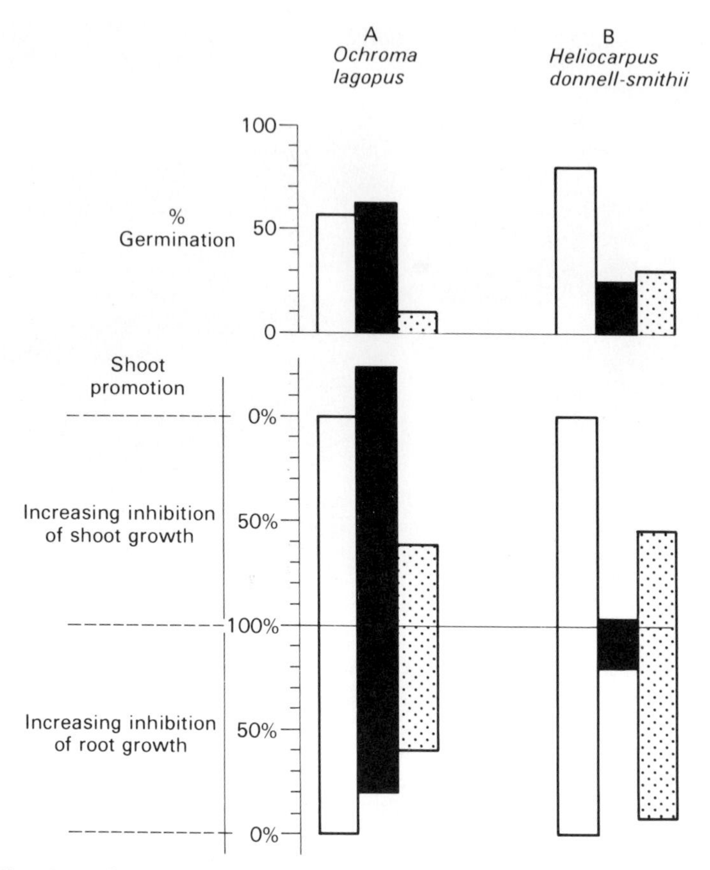

Fig. 6.9 Selective allelopathy amongst pioneer tree species in Mexico. Effects of aqueous leaf extracts on the germination and growth of seeds of (A) – *Ochroma lagopus*; and (B) – *Heliocarpus donnell-smithii*. **Open** bars = germinated in water (control); **dark** bars = germinated in aqueous extract of *Piper hispidum* leaves; **dotted** bars = germinated in aqueous extract of *Cecropia obtusifolia* leaves. (After Anaya Lang 1976.)

germination, and inhibit shoot and root growth, in the other two (Anaya Lang 1976). These are apparently not just toxicity effects, for *Ochroma* germination was not affected by one of the aqueous leaf extracts (*Piper*), whereas for *Heliocarpus* both *Cecropia* and *Piper* were inhibitory. Similarly, the *Piper* leaves were strongly inhibitory to the growth of *Heliocarpus*, but not to *Ochroma*. *Cecropia* leaves inhibited in all instances except the root growth of *Heliocarpus*. The litter and soil beneath *Shorea albida* trees have been shown experimentally to have an allelopathic effect in reducing the photosynthetic rate in the competing species *Gonostylus bancanus* (Brünig 1983). In contrast, the subtropical rain forest tree *Grevillea robusta* has been found to produce from its roots a factor toxic to its own seedlings, which may perhaps explain why this species does not grow gregariously (Webb *et al*. 1976).

One of the most important relationships in the tropical forest is that between vascular plants and fungi. Some of the direct linkages between them are genuine symbioses, especially the mycorrhizas developed jointly by tree roots and members of the fungal groups Basidiomycetes and Zygomycetes (Janos 1983). Linkages with the former group take the form of **ectomycorrhizas**, with a sheathing mantle of hyphae covering the surface of many short terminal roots (Fig. 6.10). These are a frequent feature of

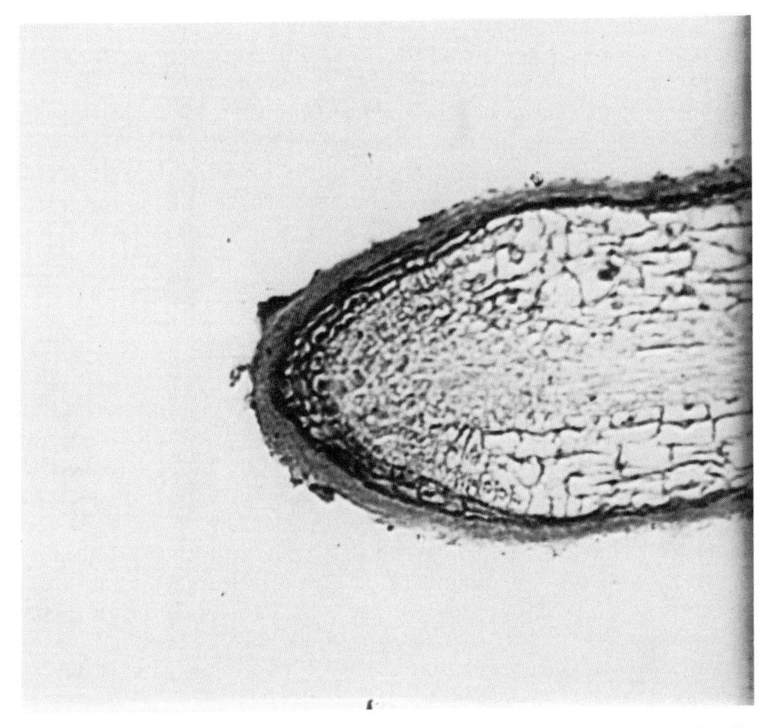

Fig. 6.10 Longitudinal section of an ectomycorrhiza in *Afzelia africana* (Caesalpiniaceae). The fungal sheath completely covers the surface of the end-root.

the root systems of trees in the Dipterocarpaceae, Fagaceae, Caesalpiniaceae and Pinaceae. More common still are the **vesicular-arbuscular mycorrhizas**, especially in lowland tropical forest, where the terminal root cells of most trees are probably capable of being infected internally by species of the zygomycetous Endogonaceae.

Both types of mycorrhizas are connected with extensive mycelia in the soil, which increase greatly the extent of biotic/abiotic contact surface within the rhizosphere. Uptake of mineral ions by the tree tends to be enhanced, especially for instance on nutrient-poor sites with immobile phosphorus. Mycorrhizal fungi may almost completely close tropical mineral cycles by their efficient uptake of mineral nutrients from the soil, and their ability to scavenge them from dying roots. It is even possible that transfer of nutrients may occur between trees of different species via fungal linkages (Janos 1983). Another important interaction between fungi and vascular plants occurs in orchids, where the fungal partner may directly recycle nutrients from litter while decomposing it. In all these cases, the heterotrophic member of the pair receives some or all of its carbohydrate supply from the autotroph.

Interesting examples of interactions between micro-organisms and plants are found in trees of the genus *Casuarina*, which often possess vesicular-arbuscular mycorrhizas, *Frankia*-nodules and dense mats of 'proteoid roots'. The mycorrhizal fungus has been shown to assist uptake of trace elements which were essential for the *Frankia*, and both organisms were found to be important to the growth of the *Casuarina*.

Some of the most striking tropical forest relationships are those between ants and plants (see also section 4.4). As well as their continual scavenging, which moves material around the entire ecosystem, ants may be specially associated with particular plant organs. For example, they pay frequent visits to extra-floral nectaries, and in some cases at least it has been shown that the pugnacious activity of these ants protects the plant from attacks by leaf-eating and plant-sucking insects (Bentley 1977*a*; *b*). Even some ferns have nectaries, which produce considerable amounts of sugars and smaller quantities of several different amino-acids (Koptur *et al.* 1982). Lipids are present in relatively large quantities in 'pearl-bodies', which are small, easily detached, non-secretory leaf emergences that are quite widely distributed in the dicotyledons (O'Dowd 1982).

Trees such as *Canthium glabriflorum* almost invariably possess an ants' nest on their bole, while they are often found within the pith of branches in *Cecropia* spp in South America, *Macaranga* spp. in South-East Asia and *Musanga cecropioides* in Africa. A still closer bond exists in 'ant-plants' such as the epiphytes *Myrmecodia* and *Hydnophytum*, where the ants live inside chambers in specialised, swollen, tuber-like stems. They contribute critically to the nutrient content of these epiphytes, and gain shelter for their colonies. In *H. formicarium* they also obtain food from the inner surfaces of the chambers (Janzen 1974). In the Costa Rican plant *Piper cenocladum*, ants of the species *Pheidole bicornis* live in hollows formed by the curved and appressed margins of the petioles, and harvest large food cells produced on the epidermis there. It has been shown by experiment that few or none of these cells were formed unless *P. bicornis* was present. The ants were also observed to remove small herbivorous insects and their eggs from the plant (Risch and Rickson 1981).

Amongst the closest relationships known are those which are obligate both for the plant and the insect. Fig wasps of the family Agaonidae are the only agencies that can pollinate certain species of *Ficus*, and cannot complete their own life-cycle except via the fig tree. Many are strictly species-specific, although there are actually a few 'cuckoo' species, which inhabit the inflorescence but do not pollinate it (Wiebes 1979). Even more surprising is the epiphytic bucket orchid *Coryanthes*, which may have ants' nests guarding it, and is pollinated by a single species, the orchid bee. This falls into the liquid in the 'bucket', and can only get out by passing first the stigma and then being loaded with fresh pollinia by the anthers. The flower attracts the male bees from as far afield as 8 km, and they need to obtain waxy chemical from the plant in order to produce a perfume which attracts the females.

Birds may also carry out pollination and insect predation roles, and various kinds of secretions may be produced by plants, presumably as attractants. Two other unusual links concern seed distribution – *Macaranga gigantea* fruits are covered by a brilliant green, sticky exudate, and many birds feeding on them have been observed to leave the tree with seeds stuck to their beaks (Taylor 1982). In contrast, the very hard seeds of *Bursera simarouba* are apparently used instead of pebbles as 'grinding stones' in the crops of seed-eating birds (Tomlinson 1980). Amongst mammals, fruit-eating bats are probably essential to the wide distribution of many tree species; and monkeys have for instance been implicated in the induced branching of *Gustavia superba* by repeated feeding on the terminal shoots (Oppenheimer and Lang 1969). These few examples may serve to illustrate the intricacy and diversity of the links and bonds that connect all the living organisms of the forest (see also Fig. 6.11).

Fig. 6.11 28 caterpillars feeding on a single leaf of a tropical plant, with 3 that are clearing the frass from the upper surface, dropped when the group were eating the younger leaves.

In essence, the resilience of tropical forest ecosystems results from the presence of many negative feed-back controls amongst its countless inter-relationships. Thus a tendency for the population size of a herbivore to increase and that of the plant producer to decrease will usually be counterbalanced; perhaps by competition for the now reduced food supply, by toxic products in the leaves, or by increased numbers of predators. Moreover, this process tends not to swing so far in the opposite direction that the herbivore's survival is threatened. Of course, extinctions of species must have occurred for as long as there have been tropical forests. But the source of the problems caused by the impact of man on the tropical forest has lain primarily in those of his activities which have set up positive feed-back systems. These typically run out of control, resulting for example in the complete elimination of a species when the size of its breeding population or habitat becomes too small, in explosive spread of weed or pest species, or in degradation of the very potential of a site to carry high forest.

The scale of man's operations is now so great, and the consequences of widespread deforestation so far-reaching, that it would be unwise to ignore the warning signs. Managers need to gain some biological understanding of the tropical forest land they are trying to manage, rather than regarding such knowledge as a luxury, or irrelevant. And ecologists and physiologists have to work harder to develop the links between science and management, some of which are discussed in Chapter 7.

6.4 Replenishment and succession

In its simplest sense, an ecosystem contains at least one living component, but tropical forests are ecosystems *par excellence*, consisting of countless numbers of living units. One of the characteristic features of living things is that they alter, through birth, growth, development, reproduction, adaptation, evolution, death and extinction. These changes take place at each level of the biosphere from the molecular up to the community and ecosystem, with spatial scales measured from nanometres to megametres, and time-scales between nanoseconds and a million years. Life scientists tend to use words that refer to the dynamic nature of their subject, including life-cycle, regeneration, ontogenesis, etc, as well as those listed above. At the ecosystem level, the terms 'forest dynamics' and 'succession' are often used to express the incessant flow of changes – the essence of tropical forest life. Recently, Oldeman (1978) coined the word silvigenesis to cover the range of processes that lead from recently established regeneration to mature climax forest that has reached homeostasis.

Succession in tropical forests is a complex, multiple process that cannot be fully understood through the traditional approaches developed in the temperate zone. For example, terms such as primary and secondary succession, or climatic and edaphic climax are accepted without question at higher latitudes, yet in the tropics they can often prove somewhat questionable. Generally speaking, tropical succession is more probabilistic and less predictable (Bazzaz and Pickett 1980), and it seems to be a non-linear process that proceeds by 'manifold routes' from pioneer phase to maturity (Ewel 1980).

There are continued disagreements as to the driving forces behind succession in tropical forests. Relatively frequently, signs can be found of 'coarse' disturbance to the ecosystem caused by the events discussed in section 3.5: severe storms, volcanic activity, landslides, etc. Such external forces, together with the influence of large herbivorous mammals and particularly man's slash-and-burn agriculture and logging, induce **exodynamic** succession. Otherwise known as exogenous or allogenic succession, this leads by way of secondary forest to a climax ecosystem, after many decades or even centuries. However, none of the empirical observations made confirm that this process actually leads to the regeneration of a similar community to that present before the perturbation. The final stages of silvigenesis tend to achieve a kind of homeostatic forest, as found for instance in the Yucatán Peninsula, where large-scale regrowth followed the decline of the Mayan culture. Despite their differences, few tropical research workers believe that tropical replenishment can be said to follow strictly the kind of successional pattern recognised in species-poor temperate forests.

Conversely, the internal forces which drive **endodynamic** (endogenous or autogenic) succession are easier to detect in the tropics (Fig. 4.4). At ground level, the commonest height for human observation (see section 4.1), the horizontal patterns of different phases can often be seen, and these become more obvious on profile diagrams and from aerial pictures. The existence of these small-scale clearings within the tropical forest led to the recognition of 'gap dynamics' (Poore 1968; Doyle 1981). The initial cause is generally the falling down of a senescent tree, or perhaps a group of trees that are pulled down together because their crowns are linked by climbers. Sometimes the formation of these small gaps may be assisted by breakage caused by wind or by a minor landslip, or by lightning strikes, as occurs in *Shorea albida* peat swamp forests in South-East Asia (Anderson 1964). Trees with asymmetric crowns may also fall over relatively easily on waterlogged sites. Another way in which small gaps form is when a tree that has perhaps been subject to environmental stress is then severely weakened or killed through fungal attack or by insects. Poore (1968) made a detailed analysis of the numbers, areas and duration of gaps in Malaysian rain forest, and showed that about one-third of the total area of the forest was occupied by recent gaps.

Oldeman (1978) introduced the useful term *chablis* to tropical forest ecology. This French word graphically describes both the fall of a tree (for whatever reason), as well as the resultant hole in the canopy (Fig. 6.12), the accumulated debris and the disturbance to the soil. In this multiple sense, *chablis* has become an important concept in silvigenesis: a basic process that triggers off a cycle of 'rejuvenation' in the forest; and one of the fundamental components of its horizontal structure (Fig. 6.13). Tree-falls produce several distinctly different light and temperature environments: (1) a crown gap, with the lower tree and shrub layers intact; (2) the epicentre in which the crown fell; and (3) periph-

Fig. 6.12 *Chablis* in a tropical forest in Borneo, with a fallen tree forming a crown gap in which regeneration is beginning.

eral zones, with a gradual transition from gap to undergrowth microclimate. In the case of an uprooted tree, local differentiation in soil conditions can also be recognised. Underneath the crown gap, mineral soil is spread around by the root system as it is torn out, large quantities of leaf litter and decaying twigs can be found at the epicentre, and along the dead trunk there will soon be a mound of decomposing organic matter releasing humus and minerals.

The biological and ecological processes which accompany a *chablis*, or succession on a larger cleared area, are frequently described in the familiar language of taxonomy. Alternatively, they may be discussed in terms of growth, life-cycles, the level at which various meristems occur, or population ecology. Observations on post-clearance succession over a period of 18 years have been reported by Wyatt-Smith (1966), while Swaine and Hall (1983) described in detail the course of succession during the first five years following clearance of an area in a Ghanaian forest. Three categories of participating trees were recognised: (a) small pioneer species, that were short-lived and unable to germinate in shade; (b) large pioneer species, found also in mature forest but unable to germinate under shade; and (c) primary or late seral species, capable of germination and establishment in shade. Tree

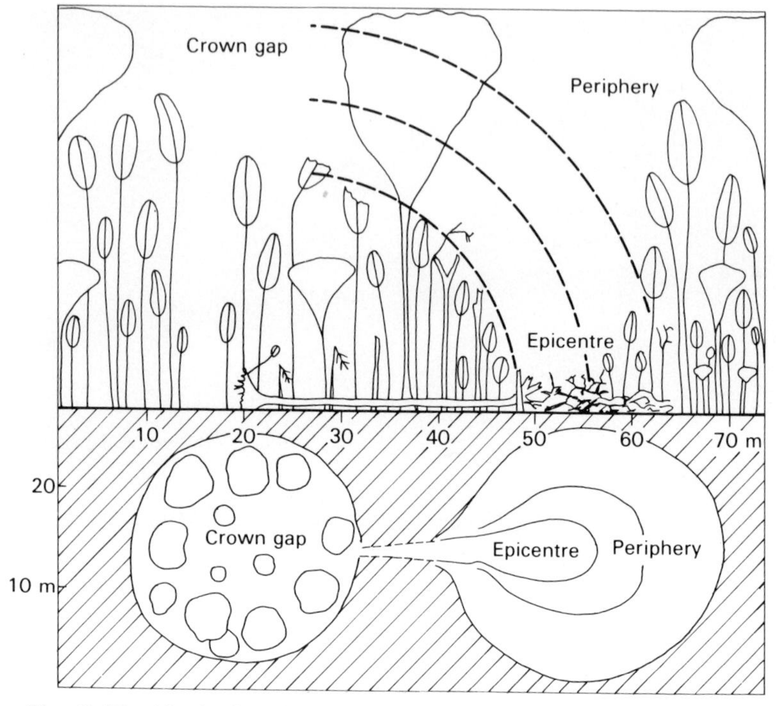

Fig. 6.13 Vertical and horizontal projections of a *chablis*. (After Oldeman 1978.)

density reached 2.5 stems per m^2 within one year from clearance, 95 per cent of which were secondary species, that is categories a + b (Fig. 6.14). After this peak, densities declined exponentially to about one tree/m^2 after five years. It was also found that 90 per cent of the total accession of secondary species was achieved in the first year. Height growth of up to 4 m/year was recorded for some secondary species, and total heights exceeded 10 m after five years, with the tallest tree being a 17 m *Trema orientalis*. By year five, the number of secondary species was declining, while about 1–2 additional primary species were being added each year.

Not only the trees, but all other life-forms are affected by a *chablis* or a more general clearance. Gómez-Pompa *et al*. (1976) have constructed a valuable model of 21 'life-cycle patterns' along a time-gradient in the successional process (Fig. 6.15). It is partly this variability of life-span and life-cycles exhibited by plants that are early immigrants, or germinate from the seed bank, which makes the mature phase of the tropical forest so unpredictable

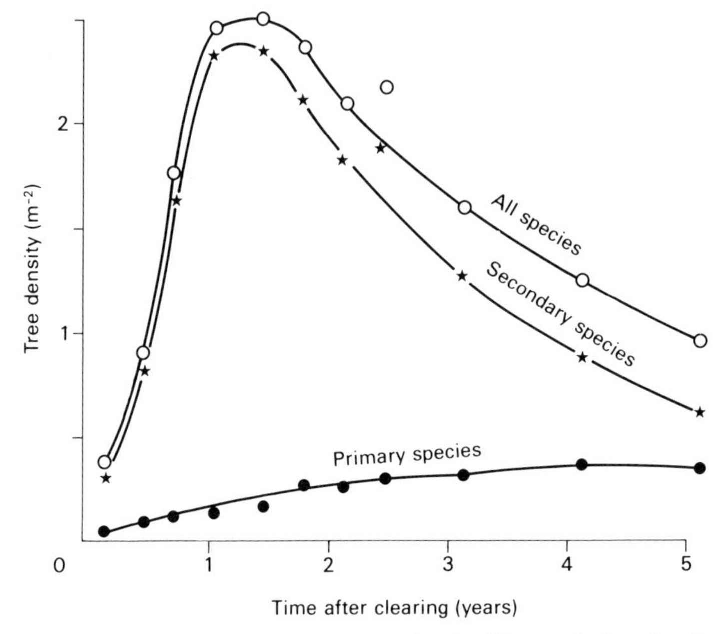

Fig. 6.14 Density of trees in a clearing in Ghana during the first 5 years of succession. (After Swaine and Hall 1983.)

and diverse. A second reason may be that many climax species have seeds which germinate promptly (see Table 5.4). Their seedlings will only have a chance of survival if a break in the canopy exists already, or occurs before they have exhausted their carbohydrate reserves (Fig. 6.16).

In temperate regions successive changes in the forest are clearly recorded in the annual growth-rings in the stem. Both the age and the growth rates of the trees, as well as their past and present inter-dependence, can be traced with a high degree of accuracy from the cut stump. In the tropics the forester and plant ecologist are generally without this reliable internal 'calendar', since the majority of trees do not form clear annual rings (see section 5.6).

Available information on the probable ages reached by tropical trees is still scarce. Data from the Indo-Malaysian dipterocarp forest suggest an average maximum age of the emergent trees of 200–250 years. Jones (1956) used several methods to estimate tree age in Nigeria, and believes that the largest trees of *Lophira alata* and *Guarea cedrata* can be 300–350 years old. The size of the tree cannot serve as a reliable indicator of its age unless successive measurements on sample plots allow reasonably

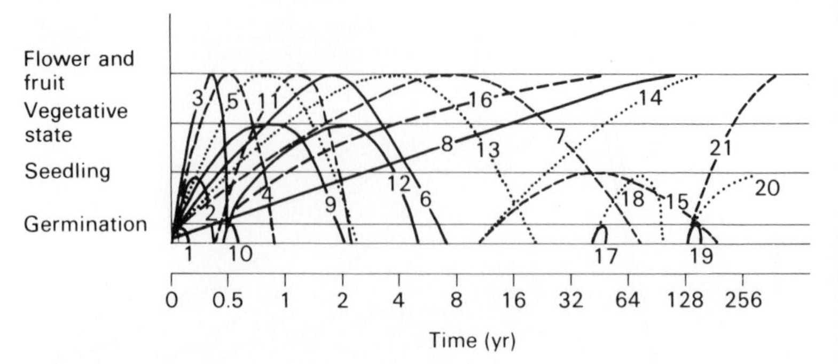

Fig. 6.15 Model showing differing patterns of life-cycles in tropical forests. 1 – germinate and die; 2 – germinate, form a few leaves, and die; 3 – finish life-span in months; 4 – annuals; 5 – biennials; 6 – life-span less than 10 yr; 7 – secondary, canopy; 8 – primary, canopy, prompt germination; 9 – vegetative only; 10 – delayed germination and die; 11 – delayed germination, annual; 12 – delayed germination, vegetative only; 13 – life-span less than 30 yr; 14 – prolonged dormancy, long life-span; 15 – prolonged dormancy, medium life-span; 16 – primary, canopy, delayed germination; 17 – prolonged dormancy, germinate and die; 18 – prolonged dormancy, remain as small plants; 19 – late stage prompt germination and die; 20 – late stage prompt germination, suppressed growth; 21 – late stage prompt germination in gaps, primary, canopy. (After Gómez-Pompa *et al.* 1976.)

accurate extrapolation. Individual emergents of similar girth may vary markedly in age, for example *Triplochiton scleroxylon* was estimated to reach the chosen diameter of 102 cm in 50 years, while *Khaya ivorensis* took about 115 years, and *Guarea thompsoni* 345 years. Estimates of tree life-span from measurements and growth simulation at La Selva, Costa Rica, varied from 52 years for *Anaxagorea crassipetala* and *Guatteria inuncta* to 442 years for *Carapa guianensis*, with a mean for the 45 tree species studied of 190 years (Lieberman *et al.* 1985). On the other hand, the life-span of pioneer species is much shorter; many American *Cecropia* spp. live only about 20 years (Hallé *et al.* 1978).

The death of old trees in tropical forests does not always occur suddenly, as a major *chablis*. The final stages of tree senescence may include the loss of parts of the root system, reductions in cambial activity and shoot elongation, stagnation and die-back of

Fig. 6.16 Natural regeneration of dipterocarp seedlings (pale, hanging leaves – probably *Shorea beccariana*) in Semongoh Forest Reserve, Sarawak, following a mast fruiting. (7 April 1982; 5½ weeks since prompt germination.)

branches, and increasing decay of heartwood in the centre of the trunk and the larger limbs. The weakened organism may be attacked by insects, fungi and bacteria, and the crown can then break off in successive pieces, followed by the progressive collapse of the trunk. While camping in the tropical forest one may hear occasional sounds of breaking timber at any time of day or night, and these are a useful reminder of the continuing changes in the forest's structure and composition.

The natural gap left by the dead tree represents a vacancy in the canopy and rhizosphere that is available to new competitors. The size of the gap may often be enlarged because of damage to the neighbouring trees from falling limbs and trunks, together with attached climbers. The environment is altered, with increased insolation reaching the forest floor, changed spectral composition of light, greater fluctuation of soil and air temperatures, more rapid release of nutrients and reduced root competition. Usually these environmental changes are quickly followed by the germination of seeds in the soil, and by promotion of the growth of stunted seedlings and saplings. In larger openings, as already described, many individuals of pioneer herbaceous and woody species generally appear rapidly and within a few years the gap has been filled with vigorous regrowth.

In this way regeneration proceeds over great areas which are not influenced by abnormal destructive factors. In regions affected by hurricanes, landslides, volcanic eruptions, fires or the sustained activity of elephants, the openings in the forest may be much larger, even covering square kilometres. This creates extensive natural secondary forests, while recent exploitation by man has made artificial secondary forests a frequent or prevailing formation in many accessible regions of the tropics.

In Central America, Budowski (1965) proposed four stages of tropical forest succession: pioneer stage, early secondary stage, late secondary stage and climax. The species which participate in the first two stages have a fairly wide distribution and within a particular tropical forest they keep re-appearing in large numbers. In section 4.5 some of the genera and species found in different regions are given. Late secondary stage species attain a considerable size and in Africa at least are often recruited from forest formations of relatively drier districts than the regenerating forest itself (see also Taylor 1960). Finally, in the climax stage, a balanced community of species is achieved, in which shade-tolerant plants grow in equilibrium with more light-demanding emergent trees. The composition depends partly on climate, soil and position in the catena, and may be regarded as a climatic climax, except where extreme soil conditions, such as impeded or excessively free drainage, or gross lack of nutrients, make one prefer to speak of an edaphic climax.

However, several difficulties arise in connection with the idea of climax stages in the tropical forest, some of which have already been mentioned. One of the most troublesome concerns the comparative rarity or even complete absence of seedlings and saplings of many of the species which form the upper tree layer. These are usually pronounced light-demanders and thus would not be expected to establish successfully in the dim light of the undergrowth. Yet, in West Africa, for instance, they may not appear even in the light phase of the smaller gaps which are quickly covered by seedlings of many other species. Results of

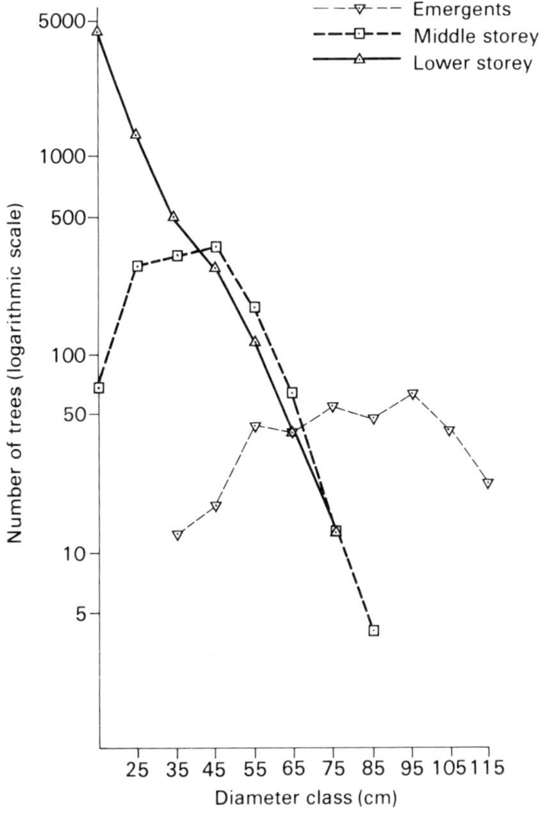

Fig. 6.17 Frequency distribution of trees with different diameters, in 3 layers of a Nigerian rain forest. (After Jones 1956.)

one very comprehensive investigation in a forest in Southern Nigeria show that the shade-tolerant species of the lower tree layer decreased logarithmically in numbers as the girth-class increased (Jones 1955; 1956). With light-demanding emergents, the position was quite different, for in almost every species there was a more or less marked deficiency of some of the middle size-classes, and sometimes even a total absence of small sizes. Trees of the middle tree layer showed an intermediate pattern of size-class distribution (see Fig. 6.17).

In order to explain this variation of the tropical forest on homogeneous sites, the mosaic or cyclical theory of regeneration is often quoted, which was first proposed by Aubréville (1938). The composition of the rain forest, particularly that of the emergent species, is thought to vary both in space and time. There are so many species that the equilibrium of the tropical forest climax can be achieved by different combinations of the available trees. Thus the forest tends to be composed of mosaics of small patches, the composition of which occurs largely at random. The reproduction of any particular patch of forest does not necessarily take place on the same spot, but may occur within other parts of the mosaic. The site of regeneration can be affected by such factors as the production of crops of viable seeds, the influence of seed distributing agencies and competition among the forest species involved. A similar concept has also been discussed by van Steenis (1958) who suggested the term 'spot-wise regeneration'.

6.5 Attempts at classification

As with other biomes, a reasonable classification of tropical forest is one of the desired aims in ecological research. Most definitions of vegetation units and synsystematic studies have been limited to a particular country, island or other small area. Only a few of them have a world-wide coverage, such as the classifications by Burtt Davy (1938) and Ellenberg and Mueller-Dombois (1967). The overall complexity of tropical forests, taxonomic difficulties, the unmanageable size of sample plots and problems of accessibility combine to restrict collection of a satisfactory number of semiquantitative *relevés*, and thus prevent application of the standard syntaxonomic procedures that are usual in the Zürich-Montpellier School. The physiognomic characters of the vegetation structure, combined with the broad features of the pertinent topography, soil and climate provide the only possible basis for a world-wide classification.

On a regional and local scale, the methods used vary too much to allow a satisfactory summary to be given. The older 'physiognomic' approach does not satisfy the usual requirements of forestry and land-use planning, and phytosociological methods have more frequently been applied in order to distinguish some detailed and economically manageable subdivisions (see Lebrun 1947; Schnell 1952; Lebrun and Gilbert 1954; Germain and Évrard 1956; Schmitz 1963; etc.). The basic vegetation unit is the association, which is further arranged in a hierarchy of units of higher rank, namely alliance, order and class. It is standard practice to give the vegetation units single or double names composed of the Latin names of the component species. A specific suffix indicates the unit's rank: for example in Katanga, Zaire, the closed tropical forests were classified by Schmitz (1963) into two classes, Mitragynetea and Strombosio-Parinarietea, the latter being subdivided as follows:

Class: Strombosio-Parinarietea
- Order: Piptadeniastro-Celtidetalia
  - Alliance: Albizio-Chrysophyllion
    - Association: Mellereto-Canarietum
    - Association: Klainedoxeto-Pterygotetum
  - Alliance: Diospyro–Entandrophragmion deleroyi
    - Association: Entandrophragmeto-Diospyretum hoyleanae
    - Association: Alchorneeto-Voacangetum africanae

Many authors, however, deny that floristic principles can be used in the classification of tropical forests, or that the concept of the association fits these conditions. The analytic phase of the phytosociological method certainly cannot be used without adequate alteration of the field procedures developed in temperate countries. In many cases classification of 'forest types' according to gross physiognomy, humidity and soil catena provides a satisfactory background for surveying, land-use and forestry management (UNESCO 1978).

The following is a tentative physiognomic-ecological classification of tropical forests, based with minor alterations on the work by Ellenberg and Mueller-Dombois (1967).

A. Tropical rain forests

(*Tropical rain forests or ombrophilous forests in the strict sense*)
Composed mostly of evergreen trees with little or no bud protection. Not particularly cold- or drought-resistant. Truly evergreen forest, that is, some individual trees may be leaf-exchanging, but

not simultaneously with all the others. Many species possess leaves with drip-tips, at least when young.

1. **Tropical lowland rain forest.** Multilayered structure composed of several tree layers and discontinuous ground-herb storey. In the upper tree layer many trees exceed 30 m height and form large buttresses. Numerous species are associated in the closed middle tree layer. Sparse undergrowth is composed mainly of regenerating trees. Palms and other unbranched trees rare, woody climbers nearly absent. Vascular epiphytes less abundant than in A2 and A3.

2. **Tropical montane rain forest.** Tree sizes markedly reduced, few species exceeding 30 m height. Tree crowns extending farther down the stem than in A1. Undergrowth abundant, often represented by tree-ferns or small palms. Abundant vascular and other epiphytes. The ground layer rich in herbs and bryophytes. Lianes present. (Corresponds most closely to the 'textbook' descriptions of the 'virgin' tropical rain forest.)
(a) Broad-leaved, the most common form.
(b) Needle-leaved or microphyllous, coniferous forest (Fig. 6.18).

Fig. 6.18 Natural stand of *Pinus merkusii* at Ban Wat Chan, Chiangmai, Thailand. The large stems and the trees on the left are pines, with dipterocarps in the undergrowth.

Fig. 6.19 Bamboo thicket composed of *Guadua* spp. in a montane rain forest of E Ecuador.

(c) Bamboo, rich in tree-grasses replacing largely the tree-ferns, pigmy trees and palms (Fig. 6.19).

3. **Tropical cloud forest.** Closed forest with numerous gaps and liane thickets. Trees often gnarled, rarely exceeding 20 m in height. Tree crowns, branches, limbs and trunks heavily burdened with epiphytes. Numerous woody lianes and herbaceous climbers. The ground extensively covered by mosses, liverworts, herbaceous ferns, *Selaginella* spp. and broad-leaved vascular herbs (Fig. 6.20).
(a) Broad-leaved, the most common form.
(b) Needle-leaved, or microphyllous, coniferous forest.

4. **Tropical alluvial forest.** Multilayered closed forest with frequent herbaceous undergrowth, palms common, and more frequent vascular epiphytes than in A1. Trees often with buttresses and stilt-roots. Numerous gaps caused by the shorter life-span of emergent trees.
(a) Riparian. Narrow strip along the river bank at the lowest level that supports forest, frequently flooded. Limited number of tree species, often much-branched. Shrubs resistant to river action. Herbaceous undergrowth nearly absent, epiphytes rare, climbers frequent.

Fig. 6.20 Cloud forest dominated by *Scalesia paniculata* on Santa Cruz, Galápagos Islands.

(b) Occasionally flooded. On relatively dry terraces along active rivers, most extensive of the A4 type. Some emergent trees achieve giant sizes of over 50 m height, with buttresses reaching 10 m height. More epiphytes than in A4(a) or A4(c). Numerous woody and herbaceous climbers.
(c) Seasonally waterlogged. Along river courses with the water stagnating for several months, sometimes twice or several times during a year. Upper tree layer rather broken and uneven in height. Trees often with stilt-roots. Middle and lower tree layer often dominated by a few species capable of vegetative reproduction. In open places palms and taller herbs occur.

5. **Tropical swamp forest.** In depressions and smaller valley bottoms with more or less permanent excessive soil moisture and poor soil aeration. Similar to A4(c), but as a rule poorer in tree species and with extensive mats of ferns or grasses forming the undergrowth. Trees less than 30 m high, frequently with stilt-roots, buttresses and pneumorrhizae, such as peg-roots and knee-roots.
(a) Broad-leaved, dominated by dicotyledonous species.

(b) Dominated by palms, but broad-leaved trees scattered in the undergrowth.

6. **Tropical peat forest.** On nutrient-poor soils covered by an accumulating organic layer. Stands lower than 20 m height except in some South-East Asian islands, composed of only a few, slow-growing, broad-leaved trees or palms. Trees are commonly equipped with pneumorrhizae and 'stilt-roots'. Few ground herbs, mostly represented by ferns.
(a) Broad-leaved, dominated by dicotyledonous plants.
(b) Dominated by palms forming pneumorrhizae and pneumathodes.

B. Tropical or subtropical evergreen seasonal forests

Mainly evergreen and leaf-exchanging trees with some bud protection. Tall species (up to 40 m) of the upper tree layer are moderately drought-resistant and shed leaves particularly during the dry season though not all simultaneously. Transitional forest between A and C.

1. Tropical or subtropical evergreen seasonal lowland forest, the commonest.
2. Tropical or subtropical evergreen seasonal montane forest. In contrast to A2 no tree-ferns are present. Evergreen shrubs and numerous climbers in the undergrowth.
3. Tropical or subtropical evergreen dry subalpine forest. Physiognomically resembling Mediterranean sclerophyllous forests. Dense stands of trees with sclerophyllous leaves, achieving a height of about 10 m. On the dark forest floor there are no ground herbs or shrubs. Epiphytes represented mostly by lichens.

C. Tropical or subtropical semi-deciduous forests

The majority of the emergent trees in the upper tree layer are quite drought-resistant and shed leaves fairly regularly during the dry period. Some of the emergents achieve a height of 40 m. Middle tree layer, pigmy trees and shrubs in the undergrowth are evergreen and may have sclerophyllous leaves. The majority of trees with some kind of bud protection, leaves without drip-tips.

1. Tropical or subtropical semi-deciduous lowland forest. Multi-layered structure and marked emergents in the upper tree layer. Some of these achieve a height of 30 m. In the undergrowth mostly tree seedlings, tree samplings and woody shrubs. Practically no epiphytes present, but both annual and perennial herbaceous climbers represented. Among the ground herbs many graminoid species.

2. Tropical or subtropical semi-deciduous mountain or cloud forest. Structures similar to C1, but trees not so tall, and covered with xerophytic epiphytes (e.g. *Tillandsia usneoides*).

D. Subtropical rain forests

Stands locally developed in Northern Australia and on Taiwan, with more pronounced temperature differences between summer and winter. Multilayered stands are similar to the tropical stands, but trees are less vigorous and allow more shrubs to grow in the understorey. The subdivisions can conform to those under A.

E. Mangrove forests

Halophilous forests in the intertidal parts of the tropical and subtropical zone (Fig. 6.21). Trees growing almost entirely in a single layer, achieving a height of up to 30 m. Trees equipped with evergreen sclerophyllous leaves and various kinds of stilt-roots, peg-roots, knee-roots and pneumathodes. Epiphytes are generally rare except for a few orchids and bromeliads, or algae, bryophytes and lichens attached to aerial roots and lower parts of stems. Ground herbs only exceptionally present (e.g. the fern *Acrostichum aureum*).

Fig. 6.21 *Avicennia germinans* with typical peg-roots at high tide near Dar-es-Salaam, Tanzania.

Many other attempts have been made to classify and map the tropical forests found in a particular country. However, the methods by which the classification units were derived is seldom given, making it generally impossible to identify a given forest stand other than by its position on the map. Hall and Swaine (1976; 1980) surveyed the closed-canopy forest actually covering or potentially occupying an area of about 80 000 km^2 of southern Ghana. The conditions that favoured such an extensive survey included considerable areas under forest reserves, good road accessibility, a relatively well known forest flora and a tradition of forest classification that had already been in operation for half a century (see Chipp 1927).

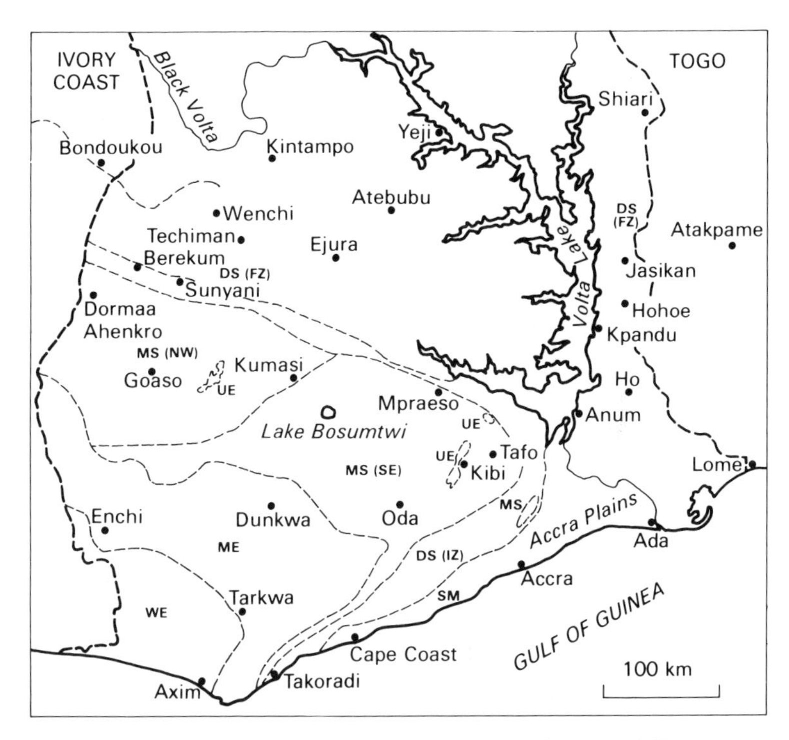

Fig. 6.22 Distribution of forest types in Ghana WE = wet evergreen; ME = moist evergreen; UE = upland evergreen; MS = moist semi-deciduous (with SE and NW sub-types); DS = dry semi-deciduous (with FZ = fire zone, and IZ = inner zone sub-types). (From Hall and Swaine 1981, using Landsat imagery to map the Volta Lake and southern forest margin (SM)

Hall and Swaine based their classification on the ordination, by a process of 'reciprocal averaging', of 749 vascular plant species, including herbs, climbers and trees. This was done for 155 plots, 25 m × 25 m, distributed more or less regularly over the whole area. Seven forest types were recognised, named and mapped (Fig. 6.22), according to their position on a two-dimensional ordination diagram. Decisions on partitioning the ordination were governed by the requirement that the forest types should be geographically coherent, and related to the environmental parameters that were also included in the analysis. Rainfall was the factor best correlated with the first axis of the ordination, while in the drier categories the forest composition appeared to be related to the underlying rock, independently of rainfall. Altitude, topography and occurrence of occasional fires were also associated with differences between forest types. This study shows that a serious attempt to improve the classification of tropical forests is not a narrow, inwardly directed undertaking, but builds on existing knowledge in an attempt to generate new ecological and biogeographical understanding.

7 Management of tropical forest land

It will be clear from the foregoing chapters that the forests of the tropics are complex and diverse, occupying and subtly modifying a variety of environments. They usually contain large numbers of species whose parallel or contrasted morphology, life-strategy and physiology combine to form ecosystems that clearly possess well integrated cycling of energy and materials, individuals and information. Many of these forests have high rates of gross and net primary production (Fig 7.1), and they contain large accumulations of dry matter and nutrients in their biomass (see sections 6.1 and 6.2).

However, as outlined in Chapter 1, attempts to manage tropical forest land have often met with failure, or only partial or temporary success. Sometimes there are unexpectedly rapid changes which prevent continued use, and if yields continue they generally represent only a fraction of what would appear possible. Why is this so? Is it impossible to obtain larger, sustained yields? Is our understanding of tropical forest biology imperfect, or has there been insufficient translation of scientific knowledge into useful guidelines? Have the managers been unwilling to listen, or the planners too prone to ignore traditional methods of utilising tropical forests? The aim of this chapter is to consider why this dichotomy between natural and managed ecosystems exists, and how botanical information could more effectively assist the practical user to make wise decisions.

A few decades ago, it might have been argued that the area under tropical forest was declining because too little was known for most of it to be managed effectively, or because this was essentially unimportant vegetation, useful only as a source of large, valuable veneer- and saw-logs. Today, however, there is certainly no shortage of scientific information being published, although there is of course much still to learn, particularly in physiology. New data continually appears on topics as diverse as ecosystem functioning and remote sensing, agroforestry methods and phenology, soil characteristics and plant/animal interactions. Interest in and funding for tropical forest research has increased, particularly in the New World. To take one example: at Barro

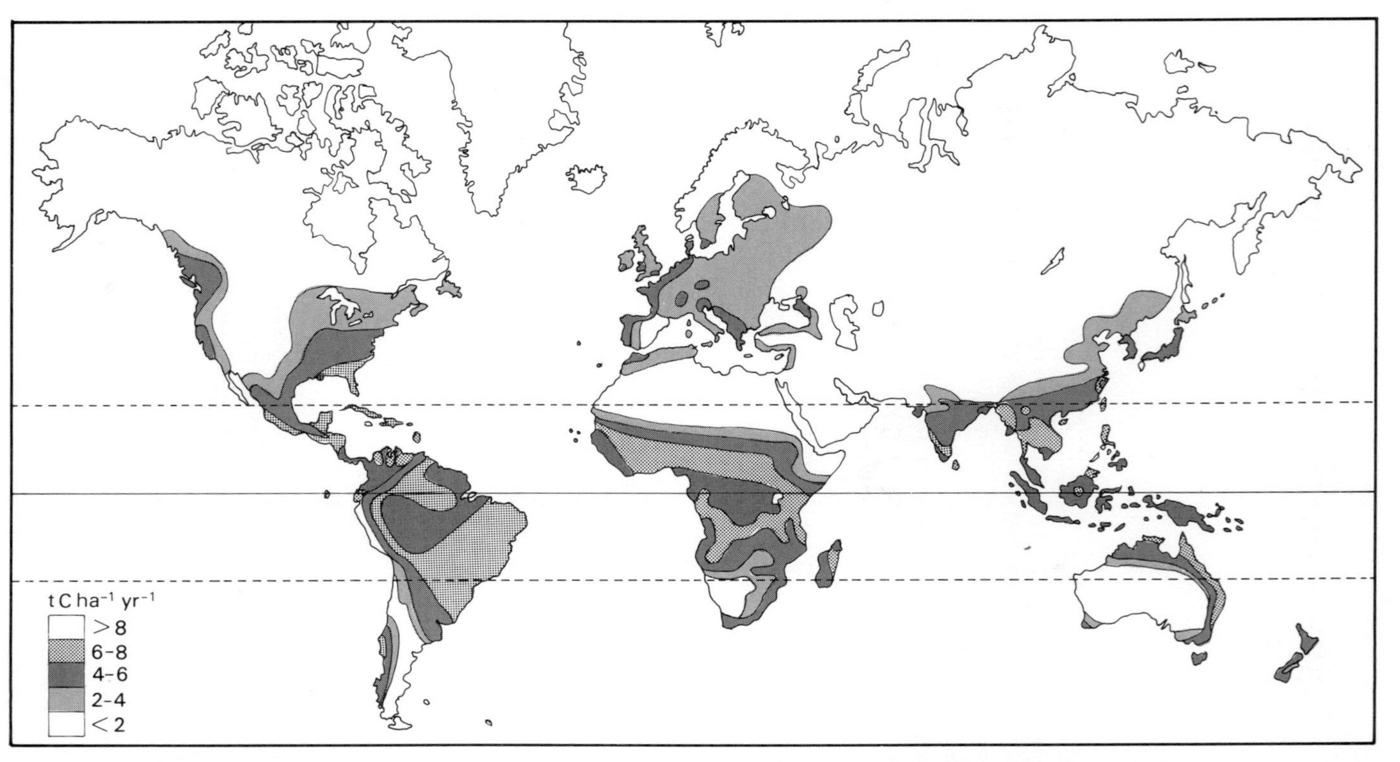

Fig. 7.1 Predicted annual fixation of carbon for the land surface of the world. (From Lieth, 1970.)

Colorado Island in the Panama Canal there is even computer-stored information on the position, size and species of every tree exceeding 20 cm dbh in a study area of 50 ha, which can be updated at frequent intervals and used to reveal many aspects of forest dynamics.

Remote sensing from satellites has revealed much about the world's tropical forests that was previously unknown (see Fig. 7.11). In Nigeria, knowledge about the percentage distribution of various types of land-use on tropical forest land has been greatly improved by detailed mapping, using Side-Looking Airborne Radar. In Malaysia, the study area and tower at Pasoh Forest Reserve, set up during the International Biological Programme, has produced valuable information on one of the few substantial blocks of lowland dipterocarp forest remaining in the area (see Fig. 2.2). Many reviews, books and proceedings of international symposia have also been published (see, e.g., Odum and Pigeon 1970; Whitmore 1975; FAO 1976; UNESCO 1978; Golley 1983*a*; Sutton *et al.* 1983; Medina *et al.* 1984), as well as many regional and local reports of projects, workshops and other studies.

Equally clearly, tropical forests are widely regarded nowadays as being of value, for a whole range of different reasons (Brünig 1977). This change has come about partly because of numerous documentary films, with a science-based narrative and superb photography, which have helped to educate people in many countries. In part, it may also reflect a growing realisation that a number of severe problems might be associated with widespread deforestation. Outstanding is this question: could the droughts that have killed many people in Ethiopia and Sudan have been partly triggered by deforestation in humid West Africa?

Elsewhere too, there are the warning signs of falling yields, shorter fallow periods and the spread of weed species. The spiralling cost of staple foods, bushmeat, firewood and everyday materials in many local markets show that there are economic stresses additional to those imposed by the vagaries of world prices, both for oil and for the traditional tropical export items. Elsewhere in the world, forests have often been over-exploited or removed, but (with some notable exceptions) they have generally been replaced by other productively managed ecosystems. However, it does not automatically follow that the same will be true in the humid tropics, where the forests fulfil multiple ecological and economic functions.

7.1 Questionable assumptions?

Before discussing some general principles underlying tropical forest land use, together with specific examples of problems and progress, it may be worthwhile pausing for a moment to consider whether there may not be certain questionable assumptions underlying our attitudes. For instance, generations of students from the tropics have been trained in colleges and universities in the temperate zones, unconsciously absorbing a set of 'axioms' about their subject which were not always relevant to circumstances in their own countries (Fosberg 1973). Providing a separate education for foresters and agriculturists, for example, tends to produce professionals expecting to manage land inalienably assigned to their own particular discipline. Those from the temperate zones who have come to work in the tropics have often had similar expectations, and have also frequently thought that efficient use of land implied large, neat blocks of even-aged monocultures, established and harvested with heavy machinery. Amongst both groups, perhaps, existing methods have often been disparaged, with the result that traditional systems that use intimate mixtures of farm crops and trees on the same piece of land, and tree-fallows, although forming the basis for tropical societies over many centuries, are now being painfully relearned as agroforestry.

Strangely enough, one of the most generally useful tenets of temperate zone forestry has often not been transferred to tropical practice. This holds that forest management is chiefly concerned with achieving (or maintaining) a situation of maximum sustained yield. Instead of being seen as a continuing national asset, to be modified and managed to produce in perpetuity, it is crucial to ask why tropical forests have come to be regarded as expendable resources, to be exploited or 'mined' until sooner or later they are gone.

One reason may be the commonly held view that managing tropical forest for sustained yields without drastically altering it is impossible or too difficult. A second could be that land-use under forestry is often considered to be 'waste', perhaps because of the prevailing temperate zone assumption that to be 'Developed' implies the wholesale clearance of forest in favour of plantation agriculture and industry. A third might be the assumption by some economists that 'natural services and goods have value only after they enter the market system or when in short supply' (NAS 1982). Indeed, even at an international symposium at Yale University in 1980, designed to bring tropical forest ecologists and economists together, declarations could be made that 'part of the tenets of our belief as economists' is that 'resources aren't resources if they're not producing anything significant'.

Thus there are powerful, largely unconscious factors favouring deforestation, in addition to the more obvious influence of the high prices paid for some of the scarcer tropical timbers, the new technologies for pulping mixed hard woods, and the requirements of many Third World countries for foreign exchange. Yet the survival of human communities must always have depended on giving somewhat greater priority to long-term goals than to short-term needs, however pressing. Principles such as 'not eating seed-corn in time of famine' have clearly been passed on through the processes of social as well as biological evolution. So if the presence of tropical trees should turn out to be important or essential to enable the land to fulfil one or more of its key continuing functions, it would be dangerously short-sighted to ignore this. Furthermore, if ways of managing forests to give substantial sustained yields were to be available, it would be a severe loss to have been deprived of these options, through the choice of guides for planning that are inherently weighted to urge the short-term gain.

No-one would recommend, for example, that in a thriving freshwater fishery ecosystem everything edible should be harvested forthwith, even the poacher perhaps appreciating the desirability of supplies still being there in future years. Management of tropical forest land may well be more complicated and difficult than that of fisheries, but these are useful parallels to bear in mind. From a study of primary and secondary production, life-cycles, and population dynamics, it is relatively easy to calculate the carrying capacity and the numbers and size of fish that can safely be harvested in perpetuity. It is necessary to take into account the likelihood of unusual ecological extremes, as these can lead to population 'crashes' which make the survivors extremely sensitive to any overfishing. For similar reasons, management of forest land should clearly start with an appreciation of its ecology.

Both forest economists and ecologists share part of the responsibility for the present confusion over tropical forest policy. Many economists, for example, have concentrated on calculating the costs of replanting deforested land. Since these are carried forward, with compound interest, to the end of the rotation, the emphasis is inevitably placed on trying to spend as little as possible. One could equally well argue, however, that when replacing a natural resource which has been lost the stress should be laid on its value continuing indefinitely. Such a view would encourage sufficient expenditure to make a success of the venture, which also holds when managing a 'normal' forest. Here stands (or trees within stands) exist at every stage from small saplings to full-sized trees, with the great financial advantage that part of each year's yield pays for the forest operations in that same year, without any interest charges.

Another problem with the assumptions behind economic calculations arises from the multiple usefulness of tropical forest (see Fig. 1.4). It is very difficult to put a figure on the value of items as diverse as maintenance of soil fertility and supply of bush-meat; or amelioration of climatic extremes and provision of logs for dug-out canoes (see Fig 7.2). Firewood and charcoal from forested areas meet a high proportion of the energy needs of people in rural parts of many tropical countries (Campos-Lopéz and Neavas-Camacho 1980; NAS 1980; 1983*a*), yet as recently as the 1970s world energy surveys were published which omitted any reference to fuel from current photosynthesis. Although in the 1980s energy from biomass (including wastes) is frequently discussed, its role is often devalued by statements that it could never meet a significant proportion of the high energy consumption in urbanised and industrialised societies. But for the

Fig. 7.2 Transporting a fishing boat, made by hollowing out a large *Triplochiton scleroxylon*, from a Ghanaian forest to the Volta Lake or the sea. The fishing industry depends upon the availability of large logs of a few suitable species.

one-third of the world's population who depend on firewood for cooking and heating (Fig. 7.3), and are faced with dwindling supplies, traditional economic analyses seem largely irrelevant. Indeed, as Roche (1979) has pointed out, 'it is not surprising that a system of forest management derived from transient and fluctuating world market demands, rather than local ecological imperatives, has proved a failure and is being abandoned'.

Fig. 7.3 Contrasted fuel needs, in a former tropical forest area of S Sudan. Because trees are scarce, the lady may have a long walk carrying her bag of charcoal. The oil tankers are waiting to cross the White Nile at Juba.

If new and relevant approaches by economists are needed, the same is true (for somewhat different reasons) of ecologists. In a sense, their problem is that they have allowed the word 'ecology' to convey misleading messages about their subject. Sometimes it seems to be an esoteric pursuit of theoretical relationships, irrelevant to practical decision-making and couched in unintelligible jargon. At other times, 'ecology' and 'conservation' appear to stand for resistance to any change, and an unreasonable desire to 'put the clock back'. The use of these terms can also look like a wish by those outside the tropics to dictate to Third World governments; or may become confused with the message of 'eco-freaks', preaching doom and gloom without suggesting realistic alternatives.

The real meaning of ecological science – studying the interrelationships between organisms and their environments in natural and managed ecosystems, therefore tends to be obscured. At best, 'ecological effects' may be included in a resource planning document, usually as an afterthought, whereas the ecology ought to form one of the central pillars of land-use. It will only do so, however, if all the relevant environmental scientists spell out clearly and positively the practical significance of their research to the development of the humid tropics.

Perhaps the most fundamental assumptions needing fresh thought concern the nature of development itself. Does it necessarily imply the wholesale abandonment of existing farming systems in favour of large plantations and ranches? Must it always entail extensive urbanisation, with industry, services and marketing highly centralised and export-orientated? Could more people actually receive a much-needed increase in their standard of living if the emphasis was placed on identifying rural development problems and opportunities? Views have been changing in this respect amongst development agencies and governments, including a distinct trend towards the involvement in projects of people living locally (Catinot 1984; Singh *et al.* 1985). Some advantages of an approach aimed at adding to local experience are that new or modified techniques and species can be thoroughly tried out before they are widely adopted; and that news of successes will tend to spread automatically, while failures may not involve a crippling loss.

The suggestion has been made that the motto for this decade should be *Trees outside the Forest* (O'Keefe 1983). Provided it is not used in an exclusive way, this neatly encapsulates the need for a far-reaching broadening of attitudes by all concerned. Adopting slogans such as 'Trees mean food' may be a useful way of bringing home the central role of perennial woody plants in the economy of managed rural ecosystems. It also implies that valid alternatives do exist or can be found, whereby tropical forest land may be managed on a sustained yield basis. How great those sustained yields and other benefits can be, and whether they can meet the reasonable expectations of expanding human populations, is one of the key questions to be addressed before the end of this century by scientists, economists and sociologists.

The solutions to the individual problems will no doubt be as diverse as the variation in topography, soils, climate and vegetation between and within each district, to say nothing of the crop preferences, hunting practices or animal husbandry systems of each of the human populations. However, by the year 2000 it might well have become clear that the greatest obstacles had been our fixed perceptions rather than a lack of biological production potential by the land. In educating the new generations, we need to change the emphasis from description of events as they now stand to analysis of why things happen the way they do (Janzen 1973).

Table 7.1 Failure and success in communication

Level of understanding reached	Possible reasons
Virtual incomprehension (apathy or hostility likely)	– Widely different assumptions (unrecognised). – Unthinking rejection of alternative views. – Foreign, or second language barrier. – Use of highly technical words; long words; abbreviations without explanation. – Speaking for too long; too fast; too quietly. – 'Waffling'; mixing with irrelevant material. – Intentional 'blinding with "science"'. – Patronising and/or servile attitudes. – Not attending.
Narrow comprehension (misunderstandings likely)	– Scope of subject not introduced. – Easy dismissal of alternative explanations. – Not seeing the wood for the trees. – Not appreciating facts relevant to the situation. – Not being aware of limitations. – Inability to adapt ideas for other circumstances.
Wide comprehension (successful communication likely)	– Using straightforward language. – Using practical examples. – Recognising limits on present knowledge. – Setting subject in its background. – Recognising mutual contribution. – Building on existing methods. – Talking in terms of the listener's interests.

Note: Good communication need not imply that the parties are agreed as to the present situation or on priorities for the future.

One of the most important steps we can all take, therefore, is to achieve better communication. The words we use may express the ideas or information we intend, or may fail to do so. Someone's partly concealed assumptions or prejudices may convey more of a message than the actual words; or his gestures, mannerisms or tone may cancel out what he is trying to say. Table 7.1 outlines some of the reasons that make communication good, poor, absent or negative – many of the problems can be avoided by the simple exercise of thinking oneself into the other person's position. In addition, there are certain words which carry strong overtones that have been superimposed on their original meaning, and thus are better avoided (Table 7.2). Good lines of communication are essential to successful management, neither an appropriate scientific background nor a wealth of practical experience being of use if we fail to pierce the comprehension barrier.

Table 7.2 Examples of words to be avoided for good communication

Word with 'loaded' implications	Suggested alternative
Development	Rural development
Ecology	Biology
Exploitation	Destructive use of resource; or utilisation
Impossible	Not advisable at the moment
Inevitable	Likely on present trends
Native	Indigenous
Peasant	Local farmer
Subsistence	Family food
Uneconomic	Not providing rapid financial return

Before leaving these basic considerations, a single example will demonstrate how important clarity can be. As has been mentioned, there is sharp debate as to whether tropical forests ought to be regarded as renewable resources. Yet the term 'non-renewable resource' can mean either: (a) one which is not being replaced by natural processes (or only at a very slow rate), such as oil or a mineral deposit; or (b) one which does replace itself, but is difficult to restore if damaged. Thus in order to allow for the paradox involved in speaking of the tropical forest as simultaneously renewable and non-renewable, it may be better to choose explanatory language such as that of Ayensu (in NAS 1980), 'we must look upon woody plants as renewable resources that, if effectively managed, could alleviate the problem not only for the present, but for posterity'.

7.2 The degrading of the forest resource

As discussed in section 2.4, man has influenced the world's tropical forests for many centuries. Sometimes his activities have had minor effects, while on other occasions they have led to profound changes in the woody vegetation, or to permanent loss of tree cover. Not all these modifications have been deleterious, of course, but problems have been sufficiently frequent and serious to justify alarm at the pace and extent of deforestation today (see section 7.5). Before discussing some of the reasons for this concern, which indeed applies not only to the humid forest zone but also to drier parts of the tropics, a brief account is needed of the various types of utilisation of forest land in the humid tropics.

There is a great diversity of ways in which man has modified, and is continuing to change tropical forests. For instance, some people still live wholly as hunter-gatherers, collecting edible parts of uncultivated plants and catching wild animals. It is usually imagined that the low population densities of such nomadic people removed so little material that the effect on the forest was negligible. However, recent evidence suggests that some groups may have caused the extinction of a few species, and considerably increased the representation of others through selection (Padoch and Vayda 1983).

Shifting cultivation has been the major system with which tropical farmers have utilised forest land, and it currently supports a population as large as that of the United States of America (UNESCO 1978: p. 469). Since it involves the cutting and often burning of patches of forest, and the growing of crops for a few years, there is clearly a substantial short-term effect on the forest. However, with small clearings and short cultivation periods, trees quickly regrow from coppice sprouts and sometimes root-suckers (see section 5.7), and from seeds arriving from the surrounding forest or present in the seed bank in the soil. When there is a low human population density, and farmed land is left for a long time before being used again, the secondary forest succession may be not unlike the colonisation of natural gaps. As was stressed in section 6.4, primary forests are not static structures, but are constantly changing, with openings formed for instance by the collapse of giant emergent trees and by other cyclical events such as floods, storms, lightning strikes, landslips and destructive insect attacks. Thus the long-term effects of stable shifting cultivation systems may again be relatively small, consisting chiefly of changes in the frequency of plant and animal species (for example, there might be fewer fire-tender trees and more fruit trees and oil-palms).

Fig. 7.4 Farm cleared for burning on a steep slope in the Andes.

However, if increased human populations lead to longer cropping and shorter fallow periods, the natural repair systems of the tropical forest can no longer quickly restore an area to high forest (Fig. 7.4). Since agricultural yields typically decrease under these circumstances, a self-accelerating, positive feed-back system has been set up, with more and more land producing smaller and smaller harvests (Fig. 7.5). Finally, the area may be abandoned as no longer able to produce enough to be worth farming (Fig. 7.10A). Thus in the absence of any perceived alternative, since people need to eat today, they may succeed in destroying the means of having food tomorrow. The changes to the original ecosystem become so profound that it is replaced by scrub (Fig. 2.11), savanna, grassland or even patches of bare land (see Fig. 2.8), with a reduced and different fauna and flora, typified by low biomass and net primary production. Weed species may predominate, such as the grass *Imperata cylindrica* (Fig. 7.9), the shrub *Lantana camara* or the dicotyledonous herb *Eupatorium odoratum*. Each year, about a million hectares of former tropical forest land is abandoned by farmers due to complete degradation (Salati and Vose, 1983).

As well as these largely unsought changes, tropical forests

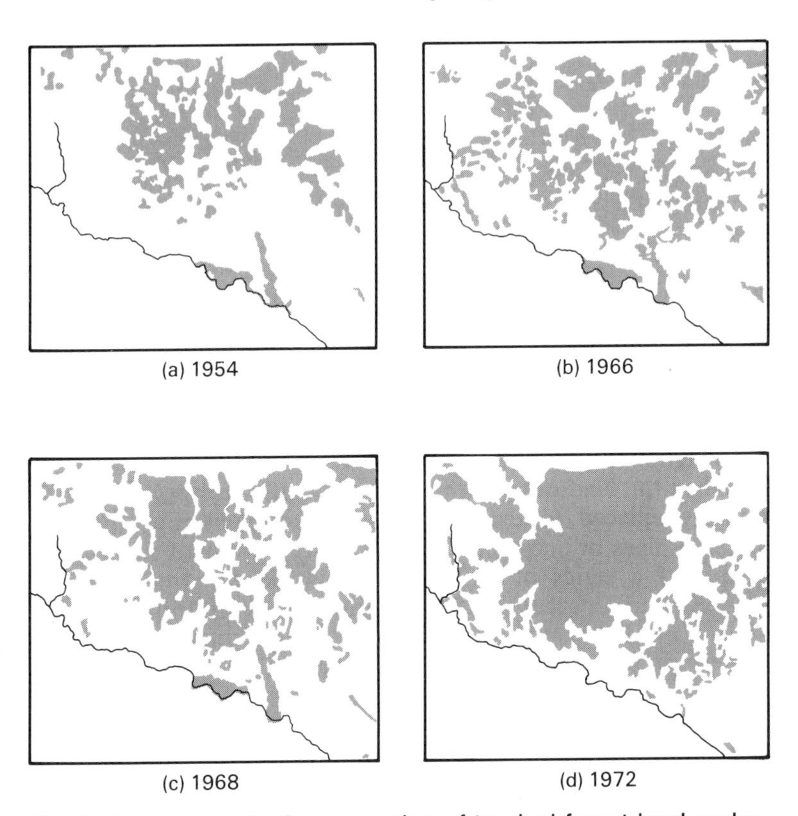

Fig. 7.5 Increase in the proportion of tropical forest land under shifting cultivation (black zones) in N Thailand during an 18-year period. Prepared using a combination of Landsat satellite images and low altitude aerial photography, the 4 maps show an exponentially increasing decline in the proportion of forested land. Formerly, the farmers selected particular soil types for farming, and used a fallow period of more than 20 years, but this area, 65 km north of Chiangmai, is under extra pressure from the increasing international trade in illicit opium. (From Miller *et al.* 1979.)

have increasingly been intentionally replaced with other vegetation. In the New World, particularly, substantial stretches of forest have been converted to grassland for extensive ranches for grazing cattle. On some of the more fertile freely drained sites in Borneo, *Imperata*-dominated farms are actually preferred (Fig. 7.9), because they are easier to clear, while large areas of inland, mountainous New Guinea consists of such age-old *lalang* farms, cropped at intervals.

Also in South-East Asia, many swamp forests have been converted into carefully managed, continuously productive rice paddies. There are also a few examples known of successful permanent-field food crop farming on freely drained sites, such as the so-called 'banana cultures' of humid East Africa, maintained over centuries by composting and manuring (Padoch and Vayda 1983). Rubber is today often grown in large blocks of even-aged monocultures established on land that has been bulldozed level, planted with a leguminous cover crop between the young trees, and then fertilised frequently for a number of years. Although such diverse agricultural practices all involve complete replacement of the forest species, their significance for the long-term fertility of the soil varies very greatly.

When the land-use is forestry, the original tree species may be entirely replaced, for example by large plantations of exotics such as teak, pines or eucalypts (see Fig. 5.24). Alternatively, smaller stands of a series of indigenous or introduced trees may be planted pure, perhaps within a larger block of tropical forest (Fig. 7.6). Here modification of the original plant and animal species may not be complete; and changes can be smaller still after selective logging, if this is carefully done, and followed for example by line-enrichment planting or by measures to encourage natural regeneration. Frequently, however, cut-over forest is simply transferred to farming or other uses, though there is increasing interest in well integrated agro forestry (see section 7.3). More drastic changes may occur in the future, because of developments in mechanisation and in mixed-species pulping technology that now permit the harvesting and conversion to paper of the boles of virtually an entire forest.

Clearly, the manager of tropical forest land needs to know what changes are likely to occur as a result of his decisions and actions. In particular, he should be able to distinguish those which lead to a degradation in the productivity and carrying capacity of the site, or a loss of environmental benefits or future supplies of useful materials. A considerable amount of information is accumulating on the many types of change which may accompany these more or less drastic alterations to tropical forest ecosystems (see, e.g., Jordan 1985; Sanchez *et al*. 1985), but it is possible to give only the briefest of outlines here.

Broadly speaking, they can be grouped into effects on soil, on atmospheric environment, and on living organisms, although of course these are all interconnected. The best-known effect of deforestation is the wholesale removal of soil by surface run-off

Fig. 7.6 Successful establishment of the valuable primary forest species *Khaya ivorensis* by line enrichment planting in a Ghanaian forest reserve. Methods such as this, with a moderate opening of the canopy, may avoid some of the problems that are encountered with large clearings.

during periods of heavy rainfall. Naturally, this is most pronounced on steep slopes (see Figs. 7.4 and 7.7), and dramatic photographs of sheet or gully erosion are often published. These effects tend to be found especially where trees have been completely removed by heavy machinery, for instance alongside new roads; or where vegetation has been cut repeatedly, as in shifting cultivation with frequent and/or extended cropping cycles. Another example is forest that has been converted to grassland and then overgrazed, where gullies as much as 120 m deep and 2 km wide have been reported in West Africa (Okigbo 1977)

Fig. 7.7 Accelerated erosion of soil on a hill rice farm, made by clearing secondary forest on steep land, 33 km E of Kuching, Sarawak (December 1985).

Equally important, though less generally appreciated, is that erosion may often be greatly accelerated above the 'background' levels normal for the particular site, without such obvious results being immediately visible. Substantial amounts of the finer soil particles are typically lost, even on land that is nearly flat or gently undulating, whenever the forest cover is depleted, simplified or removed. For example, in studies involving small plots with different levels of forest cover on a 20–25 per cent slope in an evergreen hill forest in northern Thailand (Ruangpanit 1971),

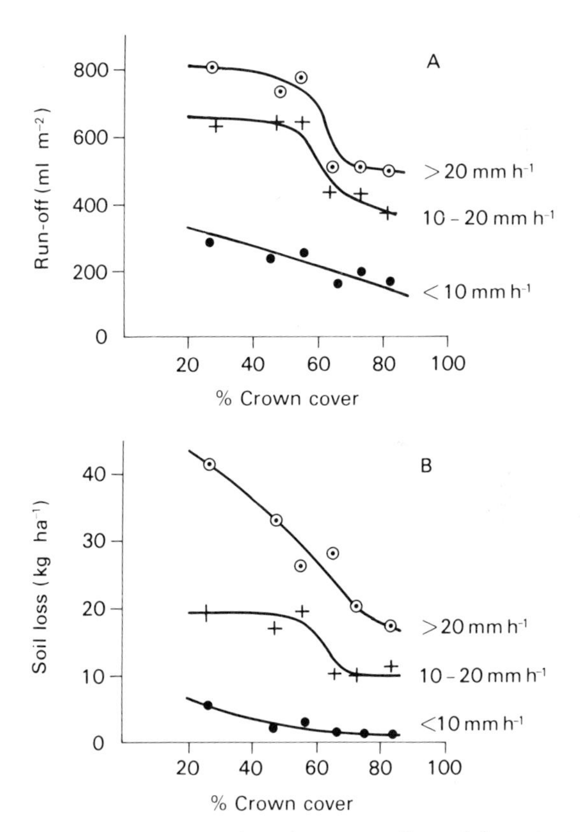

Fig. 7.8 Effects of decreasing the proportion of forest cover on (A) run-off and (B) soil loss in N Thailand. Data for experimental forest plots that had been thinned to a greater or lesser extent; and for 3 categories of rainfall intensity. (After Ruangpanit 1971; sec. UNESCO 1978.)

both run-off and loss of soil material were 50–150 per cent greater when cover was below about 70 per cent (Fig. 7.8). Soil loss also increased considerably with the intensity of rainfall, especially when it exceeded 20 mm/h, presumably because with extra kinetic energy the raindrops could wash away more soil.

It appears to be generally true that the vegetation cover is the most important of all the factors by which accelerated erosion may be controlled in the tropics. For example, annual soil loss on a 7 per cent slope in a rain forest in the Ivory Coast was 0.03 t/ha, but on bare ground it was 138 t/ha (UNESCO 1978: p. 264). The differences between average annual erosion rates under different types of cover are so large that they can best be appreciated as orders of magnitude (Table 7.3). Annual rates of erosion thus depend quite strongly on how rapidly vegetation regrows. It should also be noted that plantations that have closed canopy may not provide much cover if there are heavy storms while trees are leafless.

Table 7.3 Generalised relationships between mean annual soil erosion and the type of ground cover. (After Roose 1975, sec. UNESCO 1978: 264)

Type of ground cover	Scale of magnitude of soil erosion
Dense forest	1
Cultivation with copious mulch	1
Savanna or grassland in good condition	10
Rapidly developing cover plants	100
Early stage of plantation	100
Bare ground	1 000

Accelerated run-off and erosion also affect adjacent and more distant lower-lying ground because of the greatly increased deposit of sediment. Rather than providing a useful addition to these soils, common results are the more rapid silting up of rivers and reservoirs, and consequent flooding. Below steep slopes the soil may be covered with anything from mud-flows to banks of shingle, while red or brown coloured river water discharging into the sea shows clearly the destination of the most valuable part of the tropical forest soil.

Other important physical changes to forest soils may occur on clearing, besides the wholesale removal of part or all of it (UNESCO 1978; Greenland 1981). These depend critically on the methods used, particularly if heavy machinery is employed (see also Table 7.5). They often include changes in water storage capacity, pore size and penetrability to water and to roots, as well as in the proportions of different sized components. Organic matter content often decreases sharply when the vegetation type is changed: for instance, even when teak forests in southern India were converted to teak plantations, the organic matter content of the upper 30 cm dropped to 40 per cent of its previous value (Jose and Koshy 1972). Clearing the forest also greatly increases both the fluctuation in soil temperature and the depth to which temperatures fluctuate (see section 3.2). The incidence of higher temperatures can affect plants both directly and indirectly: for example, growth of crop plants often declines with soil temperatures above about 35 °C, and death may occur above about 40° C (Lal 1981). Water relations are also drastically altered at such high temperatures, as well as soil chemical processes, which in some cases may lead to the formation of hardened laterite layers.

Among important chemical changes in the soil, some of which stem from physical alterations following modification of vegetation, are the levels of exchangeable cations, content of organic carbon, nitrogen and micro-nutrients, liability to aluminium toxicity, and pH. It is difficult to generalise about these, though two points may be made. Firstly, as described in section 3.4, a sizeable or major proportion of the nutrients in tropical forest ecosystems is present within the large biomass rather than in the litter and soil. Harvesting the whole above-ground portions of the large trees, as is sometimes now practised for mixed hardwood pulp, is therefore particularly damaging. Secondly, particularly after burning, the soil may well be chemically more favourable for plant growth, but can then soon become less fertile – for example, because of rapid leaching. Moreover, burning volatilises some of the nitrogen and sulphur, as well as carbon (Ewel *et al.* 1981).

Alterations to the local atmospheric environment have been partly covered in the comparisons between microclimate within forest and in large clearings (see sections 3.1 to 3.3). In general, a felled site undergoes a more fluctuating set of light, temperature and water conditions than it did under forest, and is therefore less 'damped' against occasional or seasonal extremes. Areas in which the forest has been modified may also be more liable to damage by wind, or may become capable of supporting wildfires during dry weather. A vital question, touched on in section 2.2, is that widespread deforestation might perhaps alter macroclimate as well, possibly even reducing rainfall thousands of kilometres away. What if increasing aridity and frequency of severe droughts in a whole region should be clearly shown to be closely related to wholesale deforestation? This is a subject area urgently requiring research, for on it depend many people's living conditions, food supplies and survival.

The living organisms of the forest ecosystem are strongly affected by changes in land-use, in many direct and indirect ways. For instance, the removal or death of stumps, seeds and other perennating organs in the soil could reduce the numbers of particular plant species, while others (including some troublesome weeds) may become predominant (see section 3.5). If a tree species is selectively logged, its population may also decline in genetic diversity and in the proportions of individuals with inherently favourable stem form for forestry; loss of its pollinators or dispersal agents could mean it became extinct. Degradation of the fauna may similarly occur, some of the larger vertebrates being especially liable to regional extinction if their food source is removed, or if patches of remaining forest are too small to support a breeding population. Insect species formerly causing little damage may become pests (see section 4.4).

Another important change may be in soil micro-organisms, particularly those playing key roles as decomposers, and in nodulation and mycorrhizal associations which affect the availability of nitrogen and phosphorus (see sections 3.4 and 6.3). Recycling of nutrients within the ecosystem may also be sharply reduced if the numbers and diversity of tree roots declines substantially. This is because of the unique capacity of the forest root mat plus mycorrhizal mycelia to recapture virtually all the mineral ions released, even direct from the litter (Stark and Jordan 1978).

It is clear from this brief summary that changes to tropical forests often lead to far-reaching physical, chemical and biological alterations that undoubtedly degrade the capacity of the ecosystem to fulfil the needs of man (see Fig. 2.11). Two questions may therefore be asked at this point: firstly, is it possible to produce sustained yields of useful produce from tropical forest land, sufficient to support a sizeable population? Secondly, can degraded sites be restored to productivity?

Recent research indicates that both questions can be answered in the affirmative. For almost 1 000 years in the tropical forests of southern Mexico, for example, Lacandon Maya communities 'achieved what has yet to be attained in the twentieth century: the creation of a diverse long-term, stable food production system in the tropical forest biome' (Nations and Nigh 1980). Their sophisticated resource management systems, still practised in a few areas, involve the recognition and choice of certain forest and soil types as suitable for farming, while others are left unfelled. After cutting and burning, well-mixed assortments of food crops are grown for 2–5 years, and the natural regeneration of forest species is supplemented by the planting of tree crops. Both current farms and former farmland are used as hunting areas for wild mammals and birds, with the expectation that a proportion of the crops will go into this secondary food source.

Water control systems on low-lying ground, and ridged fields and terracing on slopes, indicate that Maya civilisations may have also practised some permanent field agriculture. Population densities as high as 180 km^{-2} were supported, and it appears to be more likely that they ceased to flourish because of outside influence, rather than themselves precipitating ecological collapse. A strong presumption must exist that where a form of shifting cultivation has persisted for a long time, it did so by avoiding the tendency to destroy the tropical forest (Gómez-Pompa 1980). Thus empirical science, such as the well-tried experience seen in the food production system of the Lancandon Maya 'offers a rational base for the development of modern sustained-yield tropical agroecosystems' (Nations and Nigh 1980).

Scientific experiments and recent technological advances can also contribute to the same aim, as shown for instances by the 'Yurimaguas technology' developed in part of the Amazon basin jointly by scientists from North Carolina State University and Peru. Contrary to what is regularly stated in text-books, it has been convincingly demonstrated that it is possible to maintain food crop yields without decline on certain common, freely drained soil types, where the changes following clearing the forest are primarily chemical rather than physical (Sanchez *et al.* 1982). The basis of the system is careful soil testing, addition of appropriate fertilisers and the use of crop rotations. In this way, combinations of upland rice, soybeans, maize and peanuts have been grown for 21 successive harvests over 8 years, with fluctuations but no downward trend in yields. Unfertilised control plots barely produced at all after the first three modest crops, and the mean annual yield during the period of the trial was only one eightieth of that of the fertilised plots.

When local farmers were asked to try out the systems developed in research plots, they were enthusiastic, for they found they could increase their annual yields 6–10-fold over the traditional average of about 1 t ha^{-1}. Even though the fertilisers were obtained at world market prices, transported 800 km across the Andes, and bought with a loan from an agrarian bank, family annual incomes increased fourfold. Thus there would seem to be a margin sufficient to absorb even a sharp rise in fertiliser or freight costs. Moreover, the farmer no longer has to take in fresh land every year or two, which undoubtedly relieves the pressure on the forest as well as on himself.

Fig. 7.9 *Imperata cylindrica* replacing lowland dipterocarp forest (background) in Borneo. Spreading by sexual and asexual means, this grass is tolerant of burning, and tends to prevent regeneration of woody species.

The second question posed by the often drastic changes following clearing of tropical forest is: are they irreversible? Some undoubtedly are, such as the total extinction of species. About some we do not have sufficient information, such as changes in regional macroclimate. Most aspects of degradation would probably correct themselves, so that the area would return to high forest if left to itself. However, the time span for such restoration of full biomass and the equivalent of original type of forest may sometimes be a matter of centuries. For example, in the white sand forests of the Rio Negro, Brazil, islands of bushy vegetation (*campinos*) occur, which have been traced to the activities of Indians of the Guarita subculture between about AD 800 to 1200 (Prance and Schubart 1978).

On the other hand, it is becoming clear that the restoration processes can often be greatly speeded up by appropriate management. For instance, in Sabah, Malaysia, striking success is being achieved with reafforestation of degraded land dominated by *Imperata cylindrica* (Fig. 7.9), of which there are approaching one million hectares (Keong 1983). The tropical Australian tree *Acacia mangium*, which apparently has nodules and mycorrhizal associates, grows remarkably well in spite of the poor conditions, suppresses the grass and other weeds, and may yield wood in excess of 20 m^3 ha^{-1} yr^{-1}. After a short rotation, coppice regrowth would be one option, or various other forest tree species could perhaps be re-introduced. In the following sections, other examples will be given of how the manager can indeed play a key role in correcting serious degradation, as well as preventing it.

7.3 Tropical forest science and the manager

It follows that successful management will depend on a careful blending of scientific knowledge about the functioning of tropical forest ecosystems with previous local experience in utilising such land. Managers must of course also take into account a variety of socio-economic factors in making their decisions, but these two considerations are basic to the avoidance of mistakes. The aim of this section is to discuss some of the general guidelines: how can various institutions and new technologies assist them in overcoming problems? How can the commonsense goals of stability and diversity of production come to prevail over the intense pressures for short-term exploitation of these natural resources?

The broadening of biological investigations to include the ecosystem level is very relevant here. Moving ahead from its initial, largely descriptive stage, ecology has become concerned with estimating the quantities of energy and materials moving through the system as a dynamic whole, identifying the key points of input and output, and studying the interplay of organisms and the flow of genetic information. Despite the great complexity of the many interactions present, this approach allows the manager to have before him relatively straightforward estimates of the scale of the partial processes (see Table 7.4) that underlie the overall productivity or carrying capacity of a particular site. As well as data concerning its maximum potential, which will be governed largely by environmental limitations, he could have information on its present status and the likely changes to be expected from natural or man-made perturbations.

Table 7.4 Scale of some dynamic processes in intact tropical forest ecosystems

Process	Approximate scale ($kg\ ha^{-1}\ yr^{-1}$)
Nitrogen fixation in soil, and on bark and leaf surfaces	10
Insect droppings, etc. reaching soil	10
Mineral nutrients returned to soil in litter	10^2
Leaf litter reaching soil	10^3
Soil brought to surface by earthworm casts	10^4
Net primary production of dry matter	10^4
Water evaporated by plants	10^7

Note: These figures give only an approximate order of magnitude, and can vary considerably from one locality to another.

Thus in the simpler example of a freshwater fishery ecosystem, mentioned in section 7.1, the maximum sustained yield harvest could be calculated for a given lake or fish-pond. This will depend primarily on the energy input from the sun, and the yearly photosynthetic production of the algae and other chlorophyll-containing organisms in the lake and the streams which feed it; and also the supply and turnover of mineral nutrients. The numbers, ages and sizes of fish in the projected yield would depend upon the life-histories and population dynamics of the species, taking into account year-to-year variation, disease, pollution, competition between herbivores and predation by carnivores.

Such an integrated viewpoint provides an extremely valuable tool for the manager, which is possible with all natural and managed ecosystems. For forests, there are extra problems caused by the large size, extensive underground parts and spatial variability which make it very time-consuming to obtain detailed measurements of leaf area, dry weights and chemical constituents per square metre of ground. Apart from major research investigations, therefore, much of the available data is likely to consist of fairly rough approximations, with little indication of statistical reliability. Seasonal fluctuations and species differences may not have been allowed for, and other important information about the plot of land may be unobtainable. Essentially, therefore, one is employing incomplete, theoretical models, which will be most useful to the manager when seen as general guides that require practical testing.

A forest also differs from a relatively closed system such as a lake in that many of the animal species will be free to move in and out of the area, while nutrients enter and leave by many routes. One ingenious way of dealing with this latter problem, that also introduces an experimental element into ecosystem study, has been developed at the Hubbard Brook Experimental Forest in north-east USA (Bormann and Likens 1969). Small forested watersheds were selected, with an underlying impervious layer and a single stream outlet. Simple, robust dams and automatic recording systems enable peak and average flow-rates and total water budgets to be drawn up, while regular sampling provides complete estimates of all mineral ions leaving the study area. The effects of alternative management options can be directly tested by comparing a watershed where the forest has for instance been felled with a neighbouring control plot, as well as with its previous performance. Somewhat similar approaches have been used in tropical forest, for instance at Barro Colorado Island, Panama, as well as in Thailand, and could be a useful development in other parts of the tropics.

Misunderstandings often arise over the various meanings of the word productivity. In common usage it can be synonymous with 'yield of useful produce', or alternatively with 'fertility'. In ecological emergetics, it means either the net primary production of dry matter by all the green plants of the ecosystem (gross production minus their own respiration), or the secondary production by animals. The NPP values that have been estimated (see section 6.1) indicate the high energy flows that pass through tropical forest ecosystems, and their large carbon reserves. Most alternatives to tropical forest fall far short of this 'maximum potential', except for some thriving young plantations, in which the above-ground production alone can average 35–50 t ha^{-1} yr^{-1}.

However, as C. F. Jordan (1981; 1983) has pointed out, the proportion of this high dry matter production that is allocated to stem wood is relatively lower than in temperate zone forests.

Because it seems that more is stored, eaten by insects and other animals, and dropped as small litter, tropical forest trees may often not be exceptional in their wood production. Much clearer information is certainly needed on the question of productivity useful to man (see Wadsworth 1983), including calculations of the proportion of NPP that is allocated to different parts of the plant, and assessments of the harvest index of clones, species and forests. The likelihood of high yields continuing to be achieved in subsequent rotations also needs further thought.

At this stage, it is necessary to ask who is the manager of tropical forest land, and what are the constraints and pressures acting on him (see Fig. 1.4). Broadly speaking, one may recognise eight different types of land-use and therefore of management situations:

(a) Activities of *small-holders*, growing mixtures of food and other crop plants, and raising a few domesticated animals. The family's requirements for food and materials may often be supplemented from nearby forest;
(b) Operations of *large estates*, growing extensive plantations, in the past typically of single species, run as public undertakings or by privately owned national or foreign companies;
(c) *Forest reserves* and *national parks*, in which the aims of protection of soil on slopes, research, conservation, education and amenity are often present, as well as production of wood. Run by government departments sometimes supported by international agencies, or by private companies under licence;
(d) *Grazing* and *ranching* with sizeable herds of domesticated animals, run by families, co-operative groups or private companies;
(e) *Freshwater swamp* and *mangrove forests*, often used for permanent-field rice cultivation, fishing and firewood collection;
(f) *Hunting* and *gathering*, particularly in more or less unaltered forest;
(g) *Urban* and *industrial land*, including road margins, etc;
(h) *Degraded* and *waste land*.

It would be outside the scope of this book to explore all these in detail, but clearly there is ample opportunity for clashes between people with widely differing goals, all wishing to utilise the same or adjacent areas. Taking an overall view, policy-makers have the unenviable task of looking at all the needs, trying to separate the over-riding from the important, the legitimate from the contrived, the unvoiced from the professionally promoted, the sustainable from the exploitive. To make any head-way, they have to be able to distinguish informed views from others that are too biased or prejudiced, and cope with a great diversity of sites, situations and possibilities, as well as a shortage of relevant information, trained and experienced man-power, and finance.

Perhaps the best way of tackling these difficult problems is to ensure that each region has an institute, station or university capable or carrying out multidisciplinary research and development on the rural environment. These need sufficient independence to provide advice that is the best available blend of existing knowledge and experience, not unduly influenced by sectional interests, and not tied to a single crop species. Their terms of reference would be to do fundamental and applied scientific research, and to summarise the main implications of this into straightforward, practical descriptions of the options by which each type of land user could:

(a) *maintain and improve* the fertility and productive potential of his land;
(b) *increase and diversify* the yields of useful produce he can harvest; and
(c) *contribute towards* the achievement of national goals and the protection of the global biosphere.

One example of the type of information which can be produced by combining fundamental studies with practically oriented trials is shown in Table 7.5. Traditional land-clearing methods were compared with one manual and two mechanised systems, and conventional ways of controlling weeds contrasted with 'zero tillage', in which weeds are kept down by retaining crop residues on the soil surface plus the use of herbicides (IITA 1978; 1980).

Yields from the maize/cassava crop were 3–4 times greater (in the first year) when the land was completely cleared by hand or by machine, rather than leaving some of the trees as is traditional practice. However, water loss through surface run-off increased 5–90-fold, and soil loss through erosion by 35–2 000-fold with these methods. So, for a short-lived and relatively modest return, future farming potential has been sacrificed. Indeed, with the traditional methods 20 grams of soil were lost for every kilogram of food produced, whereas 11 *kilo*grams (500 times more) were lost with the heavy treepusher and root rake method. In general, mechanised land-clearing was most harmful, though the use of a shear-blade (by an experienced driver) was considerably less so, presumably because soil and root systems were less disturbed. Zero tillage reduced run-off and erosion, without significantly

Table 7.5 Effects of methods of clearing and tillage on run-off, erosion, and yields of maize at Ibadan, Nigeria (After IITA 1980)

Land clearance method	Burning	Conventional tillage	Water lost by run-off	Soil lost by erosion	Grain yield	Ratio: soil lost/ food harvested
			(mm)	(t/ha)	(t/ha)	(kg/kg)
Traditional clearing (leaving some trees)	+	+	2.6	0.01	0.5	0.02
Complete manual clearing	+	−	16	0.4	1.6	0.23
Complete manual clearing	+	+	54	4.6	1.6	2.9
Crawler tractor (with shear blade)	+	−	86	3.8	2.0	1.9
Crawler tractor (with tree pusher and root rake)	−	−	153	15	1.4	11
Crawler tractor (with tree pusher and root rake)	−	+	250	20	1.8	11
Ratio: largest/ smallest value			95	1960	4	550

Note: Burning in traditional and shear blade treatments was in *situ* (+), while with tree pusher and root rake it was in wind-rows (−).

affecting yields, probably due in part at least to the mulching effect (see also Table 7.3).

The experiments just described were done at a large internationally financed Institute. Collaboration between pairs of research stations in the Third World and developed countries can be another useful way of generating valuable and relevant information, provided there is continuity of personnel as well as regular inter-visitation. On the extension side, it is important to remember that well-designed demonstration plots and clear exhibits have a key role to play, for, in the words of the Russian proverb, 'Better once to see than ten times to hear'. Moreover, the actual participation of farmers, foresters and other managers in field trials can do much to relate the scientific findings to their practical circumstances and problems.

A considerable shift of emphasis is indeed occurring among national and international funding agencies towards community reforestation projects, rather than single crop, capital-intensive plantation schemes. Such 'social forestry' approaches are to be welcomed, in that they are more likely to address the real needs of the region (Fernandez and Ocampo 1980; Catinot, 1984). They may also help to avoid some of the problems of land alienation and illicit timber cutting which can occur if large reserves are allocated to an 'outside' body. Nevertheless, they are by no means easy, requiring considerable efforts by all concerned to communicate effectively (see section 6.1), especially until the word gets around that the particular modification or addition to conventional practice actually works. Then they may well be replaced with different problems, such as producing sufficient improved seeds or plants, training enough advisory staff, and protecting research innovators from undue pressure, so that they can maintain momentum and respond to unforeseen problems.

Accompanying these and other far-reaching changes in thinking has been the developing interest, throughout the tropical world, in a bewildering variety of multiple land-use systems. As already mentioned, the term agroforestry implies the recognition that trees and farming go together, rather than being antipathetic (King 1979). The use of bush fallows between food crops is one long-standing example of agroforestry, while *taungya* schemes have a shorter, only partially successful history of combining the felling and temporary cropping of forest by local farmers with the establishment of plantations. Perhaps the most instructive of the traditional systems are those which utilise the same piece of ground more or less permanently, such as the village-forest-gardens of Indonesia. Far from being 'primitive', these sophisticated mixtures of useful fruit trees, palms, shrubs, food crops and domesticated animals have reached an advanced stage in the imitation of natural forest ecosystems. They occupy from 20–60 per cent of the total cultivable land of villages in Java, and together with land under rice and other staple crops they help support very high population densities, exceeding 1 000/km^2 in some rural areas (Michon 1983).

A great many interactions presumably occur within agroforestry mixtures that permit some combinations and procedures to provide multiple outlets from a piece of tropical forest land in a continuing, sustainable fashion. Plant science has already begun to build up a substantial body of knowledge and predictions which can improve our understanding of what may be possible (Huxley 1983; OTA 1984), and three examples will help to demonstrate the kind of benefits which may accrue.

A study of the indigenous tree species which are preferentially retained in traditional farming practice in southern Nigeria has shown that, besides fruit-bearing species and a few high-value timber trees, there appear to be some which are kept primarily for their soil-conserving qualities. *Acioa barteri* is one of these which is being further studied, and it has been found to cover bare ground very quickly, the foliage providing useful mulch and the plants regrowing rapidly after pruning or coppicing.

Alley-cropping is a new agroforestry system in which, for instance, five rows of maize might be grown between lines of *Leucaena leucocephala* (IITA 1980; Rachie 1983). This strongly growing bush (in the Mimosaceae) provides light shade during the dry season and may be cut back 2–3 times a year to provide a mulch which can contribute about 100 kg N ha^{-1} yr^{-1} to the soil, in addition to that added directly by decaying *Rhizobium* nodules. With such judicious admixtures of woody leguminous plants, cropping of the food plants can apparently be continued for longer periods than in pure stands. Thus there is clearly much scope for improving the soil cover, growing conditions and nutrient status of farm crops by intercropping with selected indigenous and introduced shrubs and trees. *Leucaena* can also be used to restore degraded land (see Fig. 7.10), as can *Casuarina* spp. for example, with their N-fixing nodules associated with the actinomycete *Frankia*. *Gliricidia sepium* is also proving useful (Chang and Martinez 1985), but some caution is needed with exotic pioneer species, lest they become naturalised and form undesirable weed pests, as has happened for example with *Melaleuca quinquenervia* in southern Florida (Myers 1983*c*).

A different traditional agroforestry practice involves the use of *Guazuma ulmifolia* trees scattered in pastures made for example on former evergreen seasonal forest land in Central America.

Fig. 7.10 Restoration of degraded tropical forest land by a colonising woody species with N-fixing *Rhizobium* nodules: (A) repeatedly farmed soil, with few earthworm casts and little natural colonisation: unfit for further agriculture; (B) soil under *Leucaena leucocephala*, direct sown to the plot 1½ years previously, and already 2–3 m tall, flowering and regenerating from seed. Note the many large earthworm casts. Experiment carried out by the International Institute of Tropical Agriculture, Ibadan, Nigeria.

Besides providing some soil cover, an extensive root system and shade for the domesticated animals, this freely sprouting species is regularly lopped for firewood and fodder, at a height too great for the regrowth to be browsed (Salazar and Rose 1984). A mathematical study of branching and growth provided simpler ways of estimating the firewood production per tree, and suggested the use of a 4- instead of a 3-year cutting cycle. This species is so productive that an average-sized farming family can supply all their firewood needs (approximately 5.7 t of green wood per year) from no more than about 30 trees, cutting back 7–8 each year.

The useful role of the plant scientist is much more extensive than can be suggested by a few examples, and includes the choice of species and the numbers and spatial arrangements that might be most productive in a great variety of mixed cropping situations. Rapid technological changes can also mean that simple gadgets and new techniques may allow fresh approaches to problems, and increased communication and travel enable developments to be tested in many parts of the humid tropics.

Amongst gadgets may be briefly mentioned a tool for planting seeds through a mulch or litter layer; many inexpensive types of containers for growing small trees, some of which are shaped to promote a good root system; cheap electric fencing for the control of animals and protection of nurseries; cooking stoves with improved efficiency, which can even burn sawdust; and simple biogas systems that provide a village community with a piped gas supply and a mineral nutrient sludge for crops from the bacterial decay of its organic waste materials.

New techniques range from changes in microbial technologies (Dasilva 1980) to remote sensing, including satellite imagery that can transform the mapping of forest distribution or land-use patterns (see Fig. 7.11); and from living (and thus termite-resistant) poles to support climbing crop plants (such as yams or peppers), to the use of auxins and ethylene-inducing growth substances to stimulate yields of rubber by 30–40 per cent by delaying coagulation and blocking of the latex vessels (Luckwill 1981). Genetic engineering may offer unusual new possibilities: for instance the non-fattening sweetener thaumatin, produced by *Thaumatococcus daniellii*, has now been synthesised by yeast cells, which could make possible the commercial production of this protein *in vitro* (Edens *et al.* 1984).

Returning to the tropical forest itself, the recognition that a much higher proportion of its tree species are useful than had been assumed in the past implies the need for a fresh look at the

Fig. 7.11 Part of E Sulawesi, shown in an Eros satellite image of 27 October 1972, using band 5 of the radiation spectrum. The dark areas are tropical forest, dissected by several large rivers; the white flecks are clouds (which often prevent reasonable images being obtained). Top left – the Gulf of Tomini; bottom right – Gulf of Tolo.

controversial question of managing it through silvicultural systems using natural regeneration. As forestry practice moves away from exploitive extraction towards sustained production, the importance of minimising logging damage becomes greater. In the hill dipterocarp forest of Sarawak, for example, damage to the residual stand can be reduced by half, without incurring additional costs, if a planned system is used, including directional felling of the trees (Marn and Jonkers 1982).

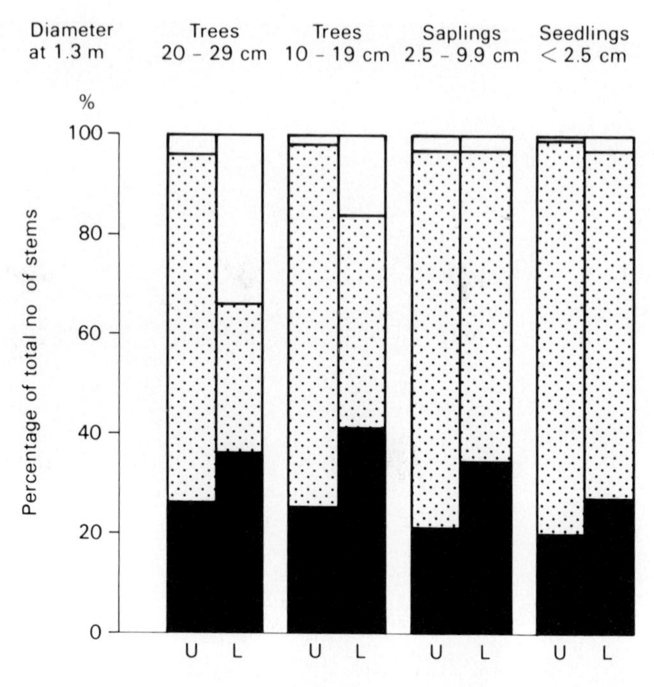

Fig. 7.12 Changes in the relative abundance of 3 groups of tree species, following logging of mixed dipterocarp forest in Sarawak. **Dark** zones – desirable climax spp.; **shaded** areas – moderately valuable climax spp.; **open** portion – heliophilous secondary spp.; U = unlogged forest; L = logged 23 years before. (After Hutchinson 1982.)

In lowland dipterocarp forest in the same part of Malaysia, it has been shown through enumerations of over 100 000 trees that selective logging, with modest opening of the canopy, had led after 23 years to a *higher* proportion of desirable timber species (Fig. 7.12). This was true in all size classes up to 30 cm diameter at 1.3 m; and the decline in the representation of secondary species amongst the smaller sizes indicates that a full stocking of young primary forest species has been restored (Hutchinson 1982). Thus it is by no means impossible for the aims of forestry and ecology to be reconciled, and for mixed tropical forests to continue to exist as stable, productive and managed ecosystems. Are there other ways in which ecology can point the way towards viable and permanent economy?

7.4 Domestication and conservation of trees

When the manager and the plant scientist realise the importance of talking together, they will also need to listen to each other. If it is pointless for managers to ignore the scientific basis of biological productivity, it is also unrealistic for scientists to argue for conservation of tropical forests unless their research is already contributing to the achievement of tangible benefits. An area where the advantages of effective collaboration could perhaps best be demonstrated has not been mentioned so far. This is the selection, multiplication and testing of genotypes of useful forest plants, and in due course the breeding of improved cultivars. A particularly attractive aspect of such genetical research is that the added value continues to accrue indefinitely.

If one considers for a moment the 'domestication' of *Zea mays*, an annual grass native to Central America, the transformations that have occurred are extraordinary. As one of the three major world cereal food crops, maize exists as varieties with an expected height of less than 1 m to more than 3 m, which can grow from latitudes 0°–50°. Some cultivars allocate much of their available photosynthates to the grain, while others are specially suited for harvesting green as forage. One estimate of the contribution that fundamental genetic research had made to the farming industry by way of breeding and selection was that it had paid for itself at the rate of 100 per cent per annum, particularly because of the large increases in food production made by farmers in the corn belts.

In sharp contrast, remarkably little is known about the genetics of tropical trees, and only a handful of species have been subjected to sustained research and improvement. Each generation of forest trees typically arises from virtually 'wild' stock that has not even had the centuries of conscious selection that successive small-holders have given to many food crops. Would it not be prudent to devote substantial sums to research on the genetics and domestication of important tree genera? After all, the returns should theoretically be considerably greater from such unimproved species than from those which must now be approaching their maximum potential. Moreover, at least in outbreeding forest trees, there is usually a wealth of genetic variation waiting to be tapped.

However, besides lack of funding, there have been a number of quite severe constraints which have greatly restricted progress in forest tree improvement. Those which are mainly biological in nature include the long time interval before trees are harvested, the breeder's difficulty in distinguishing between genetic and other sources of variation, the long delays before trees start flowering, and the unpredictable and inaccessible positions in which flowers are produced. Another problem is the smallness of the improvement to be expected from the approach often adopted, namely the taking of seed from outstanding but untested phenotypic selections, often without knowledge of the male parent. However, some of these problems can now be solved or avoided, so that in many forest tree species the potential exists for quite rapid and substantial improvement.

In its widest sense, domestication includes any conscious or unconscious act which tends to favour one particular species or genotype against another. Most forestry operations may therefore constitute a step away from 'the wild', notably in the establishment and thinning of plantations of indigenous or exotic species, recently discussed for the tropics by Evans (1982). While it is impossible to do justice to this important topic here, a few general comments are appropriate. Examples have been given in earlier sections of rapid early growth in height and diameter of various tropical forest tree species, while numerous instances of plantations are on record in which the net primary production exceeds 30 t ha^{-1} yr^{-1} and the average volume (or mean annual increment) exceeds 20 m^3 ha^{-1} yr^{-1}. Because such plantations rapidly cover the ground, they offer considerable advantages in terms of soil protection, weed suppression and the early interception of a high proportion of the incident light energy. In many ways, therefore, they may turn out to be an ideal way of using part of the tropical forest land, particularly when much biomass is needed locally – for example, as firewood or paper pulp, or when degraded areas could be restored.

Some of the disadvantages of plantations have already been mentioned, such as the accelerated soil erosion which may sometimes occur (Table 7.3). It is not yet known how general such effects may be, nor whether they could be avoided by using different methods of establishment (see Fig. 7.13 and Sanchez *et al.* 1985). It is also difficult to predict, for instance, the chance that a promising exotic species might succumb to a relatively rare ecological 'event', or an indigenous monoculture be severely attacked by a pest or disease. An example which demonstrates the need for caution is that of an unexplained die-back and death of previously healthy 30-year-old plantations of *Terminalia ivorensis* in West Africa, established with locally collected seed of this frequently planted, indigenous species (Ofosu-Asiedu and Cannon 1976). Since a risk is the product of the likelihood of an event and the consequences of it happening, managers might be well advised to diversify, rather than concentrating on single

Fig. 7.13 Load of large logs passing an area of tropical forest in which the canopy has been partly opened up to underplant with *Cedrela odorata*. (North of Kumasi, Ghana, 1965.)

species or systems. In the late 1970s, for instance, four *Pinus* spp. and one *Cupressus* sp. accounted for nearly all the conifers planted in the tropics (Wood 1979), almost invariably as pure exotic stands.

Further work is urgently needed, for instance to examine whether mixtures of tree species, perhaps with contrasting architecture or leaf characteristics, might have advantages in light interception, mineral nutrient cycling, or tolerance of stress. Meanwhile, since the planting of trees clearly needs to be greatly increased, what are the chances of this being done with genotypes more suited than the wild types to being grown in a productive, managed ecosystem?

Only the first step, choice of provenance, has so far been taken in most cases, and it is already evident that substantial differences can exist between the growth and survival of seedlings originating from various points within the natural range of a species. It is also clear that a seed origin which performs well at the native locality or in one foreign country may be the least successful in another region. This implies that genotype/site interactions can be important at the provenance level, and confirms the need for thorough local testing. However, since it is perfectly possible for individual trees of a poor provenance to perform better than the worst trees of a good one, there must be a wealth of genetic variability which is still untapped by selection at this level.

Hitherto, the next step was to select phenotypes, of well above average form and yield, and not showing any damage or disease, and propagate these vegetatively from the upper crown by grafting shoots on to seedling rootstocks. The resulting mature clones of 'plus trees' are then planted at wide spacing in a seed orchard, and whenever flowering starts they are intercrossed. When the progeny have been evaluated in trials at different sites, the below average parents are removed and any seed the orchard produces is used for establishing commercial, improved plantations.

In *Pinus*, this system has often been quite successful, although it is rather slow, and the levels of genetic gain are modest. Problems have been encountered with many other coniferous and broadleaved forest trees, including prompt and delayed graft incompatibility, absence or sparseness of flower production, 'domination' of the orchard by a few heavy pollen producing clones, and lack of synchrony in flowering times. Moreover, at least where selection of 'plus' trees is done in pure, even-aged plantations, there is now considerable doubt as to whether the largest trees will in fact give rise to higher yielding progeny. Competition studies indicate that typically the large trees are evenly arranged in a stand, rather than showing the random distribution to be expected if there were a high degree of genetic control (Ford 1976; Cannell *et al.* 1977). On the other hand, selection for differences in stem form and branching habit seem likely to be more effective, so the seed orchard and progeny test procedures may be expected to produce improvements in quality, if not particularly in quantitative aspects of yield.

Prior phenological studies can allow the selection of parent clones with compatable flowering times, and the entire management of seed orchards becomes greatly simplified when a reliable flower induction method becomes available. Amongst the many factors which appear to influence flower initiation (see sections 5.7 and 5.8), the most readily utilised is treatment with plant growth regulators (Longman 1985). Prolific male and female coning can, for instance, be regularly stimulated in trees of the Cupressaceae by injecting or spraying them with gibberellic acid (GA_3). Recommended doses are 1–5 mg GA_3 per metre of treated shoot, and up to 10 times more for trees above 3 m tall. The hormone is dissolved in a small volume of alcohol and injected into two downwardly directed holes drilled into the xylem on opposite sides of the tree (Longman *et al.* 1982; Dick and Longman 1985).

Miniaturised seed-orchards have been developed with a temperate zone cypress, in which 2–3 m tall clonal cuttings at about 1 m spacing can start to produce seed crops in five years. Seed yields can exceed 100 kg/ha, and germination percentages can be high, probably because a good pollen cloud is produced and the trees are nearer their neighbours than in traditional orchards. Juvenile as well as mature tissue will respond to GA_3, making possible a 1–2 year breeding cycle, as well as the use of seedling seed orchards. *Cupressus lusitanica*, a widely used tropical upland plantation species, and the subtropical *Callitris* spp. are most likely to respond to similar treatment.

What are the prospects for similar advances with other tropical trees? Less spectacular and reliable enhancement of coning has been achieved with a mixture of the two gibberellins GA_4 and GA_7 in the Pinaceae, including the subtropical *Pinus taeda* (Greenwood 1981), but these growth substances appear to be often inhibitory in angiosperm trees. The long-known flower-inducing technique of bark-ringing (girdling) may be the most likely treatment both to stimulate reproductive activity, and also to permit the screening of different growth regulators or other treatments for their effectiveness, within a situation in which flower initiation is more likely.

Despite these new possibilities, a time-scale of many decades to a century or more is to be expected for substantial domestication of most tropical tree species through the use of traditional methods of tree improvement. However, there is an alternative approach which has been strangely neglected by foresters: to produce planting stock directly from selected individual plants by multiplying them vegetatively, as is the normal practice with 20 or more tropical crops. The advantages and possible drawbacks of applying such direct vegetative propagation techniques to forest trees can conveniently be considered by examining an actual case-history – The West African Hardwoods Improvement Project. This was set up in the early 1970s as a collaborative venture, supported by the Nigerian Government through its Forestry Research Institute and by the United Kingdom Overseas Development Administration. Twin projects were run at Ibadan and in tropicalised glasshouses at the Institute of Terrestrial Ecology's laboratories near Edinburgh. In the 1980s it is hoped that the work will be expanded as a Regional Programme involving up to seven countries in West and Central Africa, as well as Britain and France (Leakey and Grison 1985).

Triplochiton scleroxylon is an indigenous West African forest tree which has figured prominently in timber exports, as well as playing an important role in local economies, for example the fishing industry that is based on dug-out canoes (see Fig. 7.2). It was chosen as the main species for the project because reafforestation had been severely hampered by shortage of planting stock, due to a combination of the irregular occurrence of good flowering and fruiting years, frequent incidence of insect and fungal attack on seeds, and difficulties with collection of the winged fruits and with their short viability periods (Bowen *et al.* 1977; Leakey *et al.* 1982*b*). The first aim of the projects was to attempt to establish sizeable areas of plantations with rooted cuttings. The second was to explore the possibilities of utilising clonal variation to achieve genetic improvement.

It had previously been thought that *T. scleroxylon* was a 'shy-rooting' species, but it was soon found that stem cuttings from young trees could be readily rooted, either under an automatic mist system or in a simple, shaded polythene frame with adequate watering (Howland 1975). The cuttings can be single-node or longer, but must have at least 5 cm^2 of leaf surface, and optimally about 50 cm^2. Auxin application to the clean-cut base tends to enhance both the rate of rooting and the number of roots, but is not essential. Rooting occurs between weeks 2 and 10, depending on the conditions (Leakey *et al.* 1982*a*), and with careful potting and handling the plants grow vigorously. After 3–6 months uniform batches of plants are available that are ideal for genetical or physiological research, or for planting out.

Over 60 000 juvenile rooted cuttings have been established in field trials, and have shown good early growth. Unselected clones were indistinguishable from seedlings included for comparison in some experiments. Because the identities of the randomised members of each clone were recorded, it was possible to separate the inherent component from the often large amount of variation within and between sites. Clonal differences were typically large, and where 4-plant plots were used could often be readily observed. In one trial, for example, the number of branches per metre of main stem (a useful measure of inherent branchiness) varied from a mean of five in one clone to ten in another (Fig. 7.14). Interestingly, the same total length of primary branches was found at 18 months in the tallest and in the shortest clone, suggesting that there might be inherent differences in the proportion of carbon allocated to the main stem. Does this mean that tropical foresters may one day get the opportunity to select clones for their harvest index (see Table 6.2)? This would have to include not only their competitiveness during the establishment

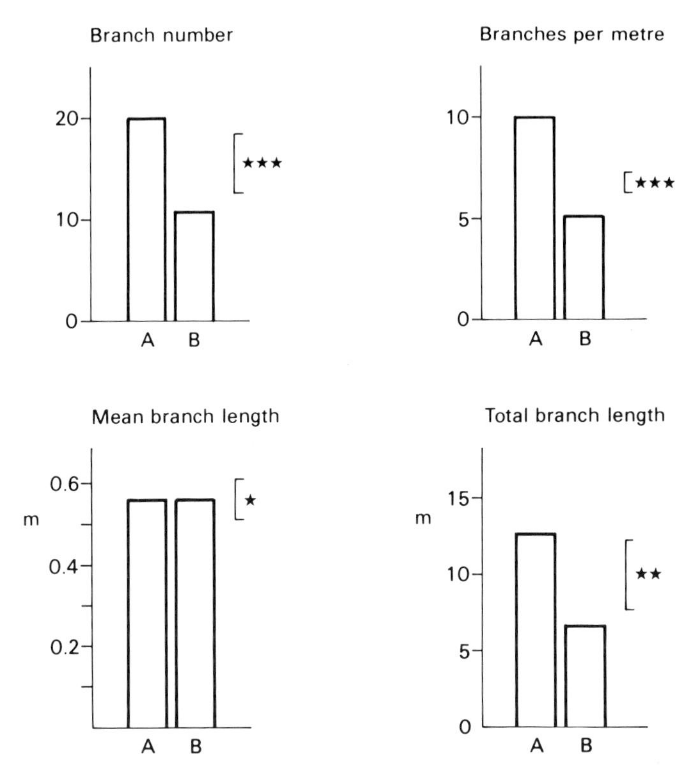

Fig. 7.14 Differences in the inherent branching habit of two clones (A and B) of *Triplochiton scleroxylon*. (After Ladipo *et al.* 1983.)

phase, but also their effectiveness after canopy closure in contributing to community production (Cannell 1979).

By working on clones of *T. scleroxylon* in controlled environments in Scotland, it was possible to show that there was a positive correlation between the number of lateral buds which grow out following the decapitation of a 10-node potted plant, and the branchiness of the same clone in the field (Ladipo *et al.* 1983). Thus it is possible to think in terms of early tests to supplement field trials, through which the least appropriate genotypes for a particular purpose could be screened off. There is also some evidence that there may be clonal differences in dry matter production, based on measurements of photosynthetic capacity adjusted to allow for respiration of the whole tree, that may be correlated with yield (Ladipo *et al.* 1984).

What are the possibilities of using such vegetation propagation methods widely in tropical afforestation? Nearly all the 75 tropical trees tested in these projects could be rooted with the same techniques used for *T. scleroxylon*. Propagation was easy in most instances, and it appears likely that this may prove to be true for the majority of the world's commercially important forest trees (Rao 1953; Opeke 1982; Leakey 1985*a*). Some of those that appear harder to root may respond to changed procedures, as occurred with *Albizia falcataria*. Leafy cuttings from young seedlings of *Nauclea diderrichii*, *Ceiba pentandra*, *Terminalia ivorensis* and *T. superba* rooted in as little as two weeks. Some species can also be rooted by planting large leafless stems directly into the ground (for example, *Bombacopsis quinata*, *Dracaena fragrans*, *Gliricidia sepium* and some *Pterocarpus* spp.), while others can be propagated from root suckers (see section 5.7). In species showing apomixis, such as *Shorea agami* and *S. ovalis* (Kaur *et al.* 1978), or polyembryony, as in mango (Chacko *et al.* 1976), genetically identical seedlings may be raised.

Amongst potential disadvantages of vegetative propagation is the possibility that the root system of a cutting might be inferior. However, when one considers the way in which seedlings planting stock is often handled, the argument for the importance of the original top-root loses much of its force. Moreover, there is the opportunity with cuttings to use auxins and good aftercare to produce well-rooted plants (see Fig. 5.17). Two other problems that also merit further investigation are plagiotropic growth habits and loss of easy rooting with time. Non-vertical growth is shown especially by cuttings from the crowns of older trees, and sometimes in species with branches morphologically distinct from mainstems, such as *Agathis*. Repeated pruning of low 'hedges' is a likely way of obtaining a ready supply of cuttings that will continue to root readily and grow orthotropically. Whatever method is adopted, it is clear that stockplant management often influences the subsequent rooting ability of the cuttings (Leakey 1983). If the stockplant has been shaded for a period, for instance, rooting may be considerably promoted over that of shoots of a similar general type that have been exposed to full light, and there is some evidence for the involvement of the red/far-red ratio.

The argument that juvenile clones are suspect because they are genetically untested fails to take account of the special opportunity they afford of direct selection in clonal trials, as well as the lack of testing of most seedling plantations. This objection is partly met if coppice shoots are propagated from the stumps of trees that have desirable phenotypes in terms of wood quality, stem form, branching, yield and tolerance of stress, pests and diseases. Not only can this approach provide a continuing supply of ideal juvenile material for cuttings, but the degree of inherent control over the selected traits will become evident in the process of replicating and 're-growing' the particular genotypes (Longman *et al.* 1978). In clonal trials of fast-growing hybrids of *Eucalyptus grandis*, *E. saligna*, *E. erophylla* and *E. pellata* in Brazil, for instance, there was a remarkably strict reproduction of the characters for which the clones had been selected (Campinhos and Ikemori 1983). Indeed, after 46 months, all but one of the 18 clones had achieved a mean height in excess of 20 m and a mean diameter at 1.3 m of more than 15 cm. It may also be noted that coppice shoots have actually been stimulated at the base of intact specimens of the temperate zone *E. ficifolia* using cytokinins (Mazalewski and Hackett 1979). Similar studies are urgently needed on indigenous genera, for example the hybrids of *Swietenia* (Whitmore and Hinojosa 1977).

The most important reservations about vegetative propagation concern the number of clones to be used. In the *Eucalyptus* example just cited, one of the clones was already outyielding the average annual volume production of the others by more than 30 per cent. Managers are likely to come under powerful commercial pressures to concentrate on one or a few 'super-clones', particularly because of the spectacular yield improvements that have been achieved by introducing new clones of rubber in South-East Asia, all derived from its extremely narrow genetic base there. Similarly oil-palm, formerly grown only from seed (Fig. 7.15), is now increasingly being grown vegetatively. Because of the obvious risks, some temperate zone countries have enacted legally-binding rules governing the minimum number of clones that must be used in forestry – an example might be 20–30 for tested, and 100 for untested clones.

Clearly, tropical foresters would be well-advised to show a measure of caution, bearing in mind how little is yet known, and how damaging would be the consequences of crop failure. Yet it is possible that well-chosen multi-clonal 'varieties' might prove as or more resilient than some seedling mixtures (Heybroek 1978; Libby 1982; 1985). Until such matters are resolved, a simple procedure might be adopted whereby several promising clones were planted (with labelled marker posts) as single clone lines at intervals amongst seedlings. This would maintain the identity of the clones, the genetic diversity of the plantation and the management options at the thinning stage. Such ideas could be rapidly tested in firewood or biomass plots grown on a short rotation, and could also be developed to fit in with species

Fig. 7.15 Extensive oil-palm plantations on former tropical forest land in S Ghana.

mixtures and agroforestry, with the aim of growing single-purpose clones in multi-purpose stands. 'The potential dangers should serve as a warning, not a roadlock to innovative and progressive programs' (Bollinger 1980).

Micropropagation of trees in aseptic culture is a specialised vegetative technique with high multiplication potential that has expanded greatly since the mid 1970s (Fig. 7.16). Considerable success has been achieved for example with callus cultures of oil-palm, which is unsuited to most other methods, having only one vegetative bud. A promising approach uses cultures in which shoot meristems are encouraged to proliferate as miniature stock-plants, from which small shoots can be transferred to a different medium to root. Plantlet formation is usually easiest when explants are taken from young trees (Leakey 1985*a*), but mature

Fig. 7.16 Micropropagation of *Khaya ivorensis* in aseptic culture at the Institute of Terrestrial Ecology, Edinburgh.

tissue from teak and other species has been successfully used (Mascarenhas *et al.* 1982). Further progress can be expected, perhaps including the formation of entirely new 'species' through protoplast fusion or other types of genetic engineering. Yet the tropical forests already contain an almost unlimited number of different genotypes, species and life-forms, so perhaps we should first concentrate on domesticating some of these.

However, the statement just made depends on the continued existence of the genetic diversity formed through long periods of evolution. Unfortunately, this is by no means certain for many species and regions (see sections 7.2 and 7.5), and so conservation of gene resources becomes an integral part of a forest tree improvement programme. These and other reasons for conservation have been discussed in numerous books and articles (for instance Frankel and Bennett 1970; Melville 1970; Burley and Styles 1975; Budowski 1976; Roche 1979; Synge 1981; Myers 1983*b*; Sutton *et al.* 1983; FAO 1985). Thus only a few points will be discussed here, despite the importance of the topic.

Conservation is indirectly assisted whenever improved cultivars or refined methods allow food, materials and energy to be produced more efficiently and on a sustained basis, because the pressure on the rest of the forest land is reduced. Direct methods include the potential for low temperature storage of the germplasm of some tropical forest plants as seeds, pollen or even tissue cultures (Harrington 1970; Withers 1980). Seedlings and cuttings representing a cross-section of the natural range of a species can be grown at several sites in gene banks, and attempts can be made to propagate rare or threatened species (see Fig. 7.18). But if many of the forest animals and plants are to be adequately protected against extermination, sizeable blocks of tropical forest will also need to be retained in each district, in order for population size to be viable.

7.5 The future of tropical forests

Unless a wholly fatalistic view is taken, the prospects for the world's tropical forest resources will depend primarily on decisions concerning their usefulness. If exploited for short-term goals they and their soils can be expected to be subject to severe impoverishment. Alternatively, if managed in one of various ways that provide sustained yields, these may individually be lower but can be enjoyed continually by subsequent generations. 'In the long term, such systems are better both ecologically and economically than systems that intensively exploit an area and yield a high profit for a few years but ultimately degrade the ecosystem's productive capacity' (C. F. Jordan 1981). Four key principles for sustaining productivity have been identified by

Jordan (1985): the maintenance of **structural diversity** in the stand, minimisation of **disturbance** of the soil and maintenance of its **organic matter**, and control over the **size and shape** of the disturbed area.

It is fortunate and timely that the extraordinary advances made in information technology are now available to help managers and planners take wise decisions. Through developments in remote sensing, data storage, computation, modelling and communications we are now in the position where we can foresee many of the results of our actions and inactions. Aerial and satellite imagery (see Fig. 7.11) are now providing reliable and up-to-date information on the present status of, and the rate of changes in the use of tropical forest land. This is clearly a basic starting point for worthwhile discussion of regional and global priorities, but used to be the subject of acrimonious debate, absorbing energy that could be employed more constructively elsewhere.

Estimates of the area under tropical forest vary from about 8 to 12 million km^2 which is approximately 40 per cent of the total area once occupied (Sommer 1976; FAO 1981; OTA 1984). There is very large variation between regions, with some areas (for example in the Amazon basin, Sumatra and Cameroon) having a higher proportion of relatively undisturbed forest, while in others (such as parts of the Caribbean, Java and East Africa) hardly any remains. To some extent this of course reflects human population levels, though it is worth noting that even in the densely inhabited island of Singapore some small areas of tropical forest have been retained. There may also be considerable variation within a region: for example, some countries of West and Central Africa still have a high proportion of productive forest, while others have very little left (Fig. 7.17).

Part of the confusion over how fast tropical forests are disappearing depends on the criteria used. **Deforestation** has now been defined as conversion of forest to land uses or wasteland having a tree-cover of less than 10 per cent (OTA 1984). On this basis about 0.5 per cent of the remaining forest disappears each year, at approximately the same rate in tropical America, Africa and Asia. However, these overall figures underestimate the scale of the problem, since the averages are strongly influenced by the presence of a few countries with very large forest areas and relatively low rates of loss. A quarter of the nations involved already have less than 0.05 ha per head of population, so clearly deforestation must here be slowing down. More than half the rest have annual deforestation rates between 1 and 6.5 per cent.

In addition, other tropical forests are suffering **degradation**:

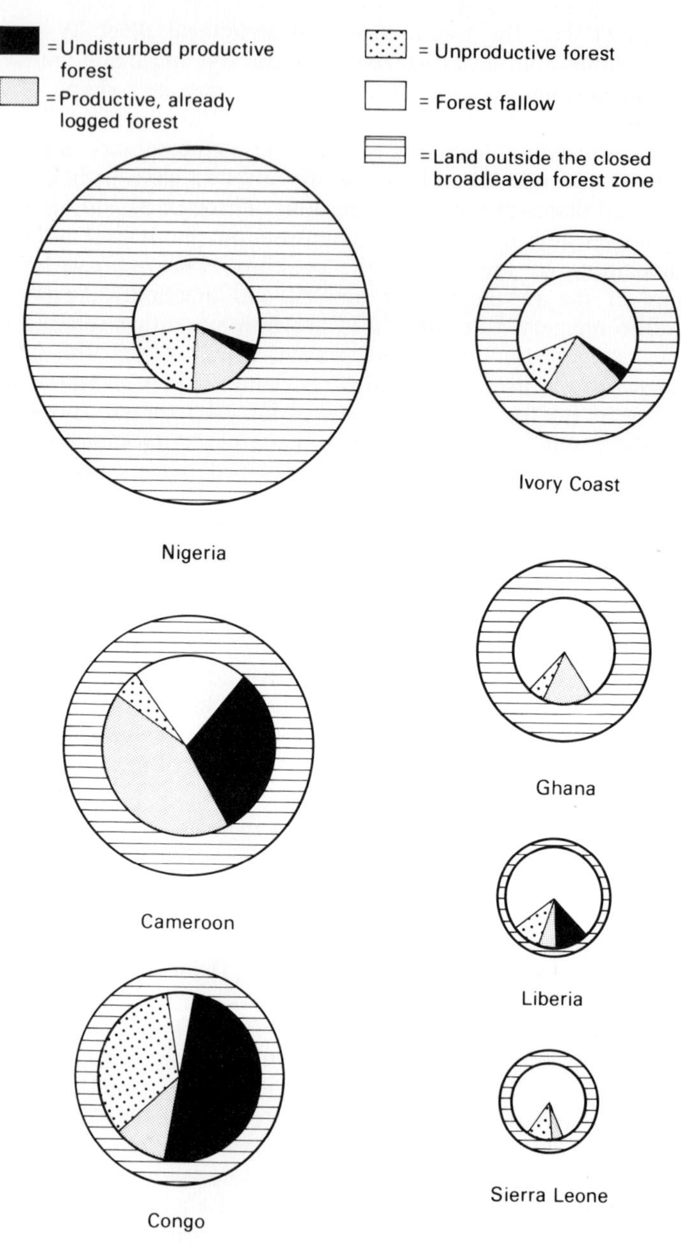

Fig. 7.17 Differences between countries in W and C Africa in the area of closed, broadleaved forest, and the extent of its conversion.

that is the loss of productive potential lasting at least a few decades, though they are still classified as forest. They may decline for physical reasons such as accelerated soil erosion, chemical processes such as loss of mineral nutrients from the ecosystem, or by biological changes such as genetic depletion or extinction of valuable species. The extent of alterations such as these is much harder to express in quantitative terms, but it is likely that **conversion** of tropical forests (= deforestation + degradation) is running at over 11 million ha a year. It is perhaps easier to comprehend this figure if it is expressed as more than 20 ha a minute.

Projections for the year 2000, based on extrapolations of current rates, suggest that the world's undisturbed tropical forests may by then have declined to about 700 million ha (OTA 1984). Other estimates are more pessimistic (Myers 1983*a*), but it is also possible that the greatly increased awareness of the importance of tropical trees may persuade people that slowing the rates of conversion is in their country's interests. Similarly, they may seek to increase reafforestation by plantations, which in West Africa for instance currently replace less than 3 per cent of the forest lost. Indeed, the total area of plantations made on tropical forest land throughout the world up to 1980 only accounts for a single year's loss.

When assessing the probable results of changes of this magnitude, it is customary nowadays to include an environmental impact assessment. This may involve worst-case analyses as well as providing estimates as to the most likely consequences, and the range of effects which may be expected. In any such study, it would be important not to overlook the influence that global conversion of tropical forests may be having on the biosphere as a whole. Such information may sometimes be imprecise, but that is categorically not a reason for dismissing it as unimportant. Indeed, societies generally make a distinction between what someone may do inside the house, but for the public good not be allowed to do outside it. The same should perhaps apply to the regulation of activities which affect what we may think of as the 'global village'.

As long ago as 1970, L. R. Brown wrote:

What is clearly needed today is a co-operative effort – more specifically, a world environmental agency – to monitor, investigate, and regulate man's interventions in the environment, including those made in his quest for more food. Since many of his efforts to enlarge his food supply have a global impact, they can only be dealt with in the context of a global institution. The health of the biosphere can no longer be separated from our modes of political organization. Whatever measures are taken there is growing doubt that the agricultural ecosystem will be able to accommodate both the anticipated increase of the human population to seven billion by the end of the century and the universal desire of the world's hungry for a better diet. The central question is no longer 'Can we produce enough food?' but 'What are the environmental consequences of attempting to do so?'

The effects of tropical forest conversion that would most obviously be of universal concern are the predictions that have been made about the global atmosphere and climate. For instance, carbon dioxide concentrations have increased by some 15–25 per cent since around 1800, when the Industrial Revolution began. One reason for this is thought to be the burning of fossil fuels, currently estimated at about 5×10^9 t of carbon a year. However, approximately 2×10^9 t/year is thought to be due to the release of fixed carbon as forests are replaced by ecosystems with a lower biomass and soil organic matter (Woodwell *et al.* 1983). Scientists' disquiet arises from what has been termed the 'greenhouse' effect: carbon dioxide molecules transmit more incoming visible light than outgoing longer-wave heat, thus altering the net radiation balance at the Earth's surface. There is a lively debate about how great an effect this may be having on world climates, but one estimate is that average annual global surface temperatures could increase by as much as 1–5 °C if carbon dioxide levels were doubled.

A doubling of the original concentration of 0.03 per cent is predicted between the years 2020 and 2050, without taking into account the probable positive feed-back of a temperature rise causing respiration to increase more than photosynthesis. Overall temperature changes of this magnitude would have profound effects on both crop plants and natural vegetation, by causing a general shift of climatic zones. Sea-level could also rise by perhaps 1.5 m in a century, through the partial melting of polar ice. However these trends are by no means inexorable, as shown by the sharp reductions in fossil fuel consumption that occurred around 1980. The alternative of burning more firewood makes sense in this context, since most of its carbon would in any case eventually be released by decomposers. A vigorous programme of afforestation could significantly help towards the stabilisation of atmospheric CO_2 concentrations by locking up more carbon in biomass (Woodwell *et al.* 1983).

Although oxygen concentrations may have declined slightly for the same reasons, the change is biologically insignificant in relation to the level of 20 per cent in the atmosphere. There is a possibility however, already mentioned in section 2.2, that deforestation of a whole region may influence water vapour cycling into the atmosphere in such a way as to decrease rainfall. If this is confirmed with new techniques such as continuously recording weather satellites, computer modelling and isotope studies, it would have profound significance for land management in the humid tropics, and perhaps also in adjacent regions subject to severe droughts.

Another question of global concern is the enhanced rate of extinction and genetic depletion of animal and plant species. This has come into prominence through publications such as the red data books, and is essentially a recognition of the many interrelationships, some of them obligate, which exist in tropical forest ecosystems. With current rates of forest conversion, it appears likely that a number of species are already disappearing from the biosphere for good each year, some of them still unnamed and unknown to science. Genetic depletion is already recognised as being important in tropical crops such as rubber and cocoa, but is also of relevance to the domestication of other tree species (see section 7.4), particularly if they have a limited

Fig. 7.18 Open flower and flower buds of the endangered species *Vateriopsis seychellarum*, produced at a height of 0.7 m in a potted plant at the Institute of Terrestrial Ecology, Edinburgh.

distribution. The endemic species *Vateriopsis seychellarum*, for example, existed in 1975 only as seven trees on the Seychelles Islands, plus a few plants kept in glasshouses (Fig. 7.18).

Irreversible loss of potentially valuable material is serious, particularly because the frequency of useful species is so high. For example, over 100 different kinds of uses have been recorded for forest trees in Ghana, ranging from fibres to fodder, and from spices to kitchen utensils (Irvine 1961). Such a wide range is unfamiliar to temperate zone foresters, who have often been trained to think only in terms of timber and paper pulp. The domestication of tropical tree species might help solve problems of food shortage, directly as well as indirectly. For example, the fruits of the New World palm known as pejibaye (*Guilielma gasipaes*) are probably the most balanced of all tropical foods, containing carbohydrates, protein and oil in addition to minerals and vitamins (NAS 1975). Twice as much carbohydrate and protein per hectare can be produced as with maize, and the fruits can also be dried and kept for six months or more. In Nigeria, at least 50 tree species have been identified as being important food sources (Okafor 1980) and 40 have been reported in East Africa (Flores Rodas 1983). A single Malaysian village may contain more than 20 fruit tree species in cultivation, some of which have undergone the early stages of domestication (Whitmore 1975).

The medicinal value of plants is often underestimated by temperate zone biologists, despite the fact that this was the original reason for their own floras, herbaria and botanical gardens. Tropical species are particularly rich sources of chemicals of pharmacological interest: for instance, details of over 150 plants (mostly trees and shrubs) used in traditional medicine have been recorded for North-East Tanzania (Hedberg *et al.* 1982–83). Moreover, 25 Mexican plants that were without known local medical use all showed positive reactions to at least one of eight pharmacological tests applied, while in a larger trial of 258 species subjected to 12 tests, 99 per cent showed positive responses and almost half had reactions that justified further study (Gómez-Pompa, 1980). Some of the biosynthetic pathways in plant cells producing such chemicals that have evolved, often apparently as protective agents against herbivores, might prove more appropriate than large energy-requiring factories for the manufacture of specific drugs (Marzola and Bartholomew 1979).

It is well known that the timber from particular tropical trees is suitable for specific purposes, such as teak and *Chlorophora* spp for laboratory benches, or balsa for life rafts and model aeroplanes. Less familiar are the species which yield flat boards from their plank buttresses, harbour piles that are resistant to rotting and insect attack, and self-insulated electric fencing posts. Timber is likely to remain one of the world's major construction materials, and in many tropical countries more and more tree species are being utilised as shortages of the first-choice timbers occur.

Less specificity is usual in the many instances where wood is used as a bulk material. The importance of firewood and charcoal has already been stressed, and indeed they together account for 86 per cent of all wood used in the Third World (NAS 1980). With increasing shortage of fuel, especially near large towns, emphasis needs to be placed on biomass studies to determine the most effective spacings and rotations for coppiced firewood plantations, and the mineral nutrition inputs needed. Genotypes suited to firewood production when trees are grown in plantations, along boundaries or in agroforestry situations, need to be found. Biomass is also a major consideration when trees are grown for pulp or particle-board, and the same would apply to other systems where wood acts as a feedstock for chemical (Goldstein 1980) or microbial technologies (Dasilva 1980), or where foliage is used to supplement wastes in village biogas generators.

As the demands on land become more intensive, one way of relieving the pressure is by utilising multi-purpose species. *Brosimum alicastrum* represents the ultimate in agroforestry – a tree that provides food, materials and soil cover (Gómez-Pompa 1980; Peters and Pardo-Tejeda 1982). The edible fruits are produced annually at 50–75 kg/tree, the seeds containing around 10 per cent of protein, including the aminoacid tryptophan, which is often poorly represented in local diets. Cattle can be fed by lopping up to half the trees' crowns for forage, or by allowing them to graze freely on seeds, seedlings and freshly fallen leaves. The tree also produces an easily worked timber, gums, medicines, beverages and firewood. Not surprisingly, this species is thought to have been an important element in Mayan civilisations (Peters 1983).

Finally, the role of trees in ameliorating living conditions in the human environment, whether in a large city or a small village, is a useful reminder that we should not take them for granted. In a sense, each page of this book has been concerned with a facet of the vital part which tropical forest trees play in the ecosystems which support our societies. More, and not less of them is the prescription gleaned from historical experience, from present-day requirements for food, energy and materials, and from expectations for a stable or improving future. The conclusion which has been reached is that there are indeed many ways in which tropical forest land can provide diverse goods and services in perpetuity.

What overall guidelines can be offered to scientists, users, managers and planners in their tasks of avoiding mistakes, correcting misinterpretations, encouraging debate and generally promoting the growing of woody plants? Scientific research needs to be multi-disciplinary, exploratory and sustained by long-term links between institutions. Its main concerns should be related to the problems and potentials of tropical forest land management. The development of new cultivars and appropriate technologies, tested under local conditions, can complement existing, well-tried species and methods. Clear explanations and good communication are especially important when blending applied science with past experience, or the innovative breakthrough with the dictates of commonsense. Many people still need to be educated to recognise that trees produce many different types of goods and benefits that make them an integral part of rural economies.

Perhaps the best way of encapsulating the present and future status of tropical forests is to quote the Asante proverb, 'Wo redi wo nkokɔ nkosua no, na wo redi wo nkokɔ-mba', which means

Fig. 7.19 *Ficus* trees taking over the ruins of the temples at Angkor, Kampuchea – a warning for all users of tropical forest land?

that while you are eating your hen's eggs, you are eating your chicken. In sayings of this kind, not only in Ghana, but in many parts of the world, past generations have expressed the fundamental requirement to restrain people from eating the seed-corn out of greed or when crops failed, without which their communities would not have survived (Fig. 7.19).

Homo sapiens has learnt how to travel in space and return to the Earth, simultaneously transmitting pictures of this enterprise to any point on its surface. He has refined his destructive capacity to a level where the entire tropical forest biome is considered by the international scientific community to be under constant threat of death through the cold and the dark of a 'nuclear winter' (SCOPE 1985; 1986). Just a fraction of the skills, effort and expenditure that are allocated to such activities could transform the prospects of survival for many of the inhabitants of the tropical world. Learning how to manage forest land wisely could be the clue to providing each one with sufficient food, a clean water supply, shelter and a future, not only for themselves, but also for the children and grandchildren.

References

Ab'Sáber A N 1982 The paleoclimate and paleoecology of Brazilian Amazonia. In Prance G T (ed) *Biological Diversification in the Tropics*. Columbia Univ Press, New York, pp 41–59

Addicott F T 1982 *Abscission*. Univ Calif Press, Berkeley

Ahn P 1961 Soils of the lower Tano Basin, south-western Ghana *Soil and Land-use Survey Mem* **2** Govt Printer, Accra, Ghana

Ahn P 1970 *West African Soils*. Oxford Univ Press

Akkermans A D L, Abdulkadir R S, Trinick M J 1978 N_2-fixing root nodules in Ulmaceae: *Parasponia*. *Plant and Soil* **49**: 711–15

Allee W C 1926 Measurement of environmental factors in the tropical rain-forest of Panama. *Ecology* **7**: 273–302

Altman P L, Dittmer D S (eds) 1973 *Biology Data Book* (2nd edn). Fed Amer Socs Exp Biol, Bethesda, Maryland

Aluko A P, Aduayi E A 1983 Response of forest tree seedlings (*Terminalia ivorensis*) to varying levels of nitrogen and phosphorus fertilisers. *J Plant Nutr* **6**: 219–38

Alvim P de T 1964 Tree growth periodicity in tropical climates. In Zimmermann M H (ed) *Formation of Wood in Forest Trees*. Academic Press, New York, pp 479–95

Alvim P de T, Grangier A, Jr 1965 Influência do fotoperiodo no lançamento do cacueiro. *Centr Pesq Cacau Itabuna Brasil, Ann Rep 1964*, p 24

Anaya Lang A L 1976 Consideraciones el potencial alelopático de la vegetación secundaria. In Gómez-Pompa A, Vázques-Yanes C, del Amo Rodríguez S, Butanda Cervera A (eds) *Investigaciones de Selvas Altas en Veracruz, Mexico*. Compañia Editorial Continental S A, Mexico, pp 428–46

Anderson J A R 1964 Observations on climatic damage in peat swamp forest in Sarawak. *Commonw Forestry Rev* **43**: 145–55

Anderson J M, Swift M J 1983 Decomposition in tropical forest. In Sutton S L, Whitmore T C, Chadwick A C (eds) *Tropical Rain Forest: Ecology and Management*. Blackwell, pp 287–309

Ashton P S 1969 Speciation among tropical forest trees. Some deductions in the light of recent evidence. *Biol J Linnean Soc* **1**: 155–96

Aubréville A 1938 La forêt coloniale: les forêts de l'Afrique occidentale française. *Ann Acad Sci Colon, Paris* **9**: 1–245

Aubréville A 1949 *Contribution à la paléohistoire des forêts de l'Afrique tropicale*. Paris

Augspurger C K 1982 A cue for synchronous flowering. In Leigh R G, Jr, Rand A S, Windsor D M (eds) *The Ecology of a Tropical Forest – Seasonal Rhythms and Longterm Changes*. Smithsonian Inst, Washington, DC, pp 134–50

Augspurger C K 1983 Phenology, flowering synchrony, and fruit set of six neotropical shrubs. *Biotropica* **15**: 257–67

Ayensu E S 1974 Plant and bat interactions in West Africa. *Ann Missouri Bot Gard* **61**: 702–24

Baker H G 1965 The evolution of the cultivated kapok tree: a probable West African product. In Brokensha D (ed) *Ecology and Economic Development in Tropical Africa*. Univ California, Berkeley, pp 185–216

Baker H G, Bawa K S, Frankie G W, Opler P A 1983 Reproductive biology of plants in tropical forests. In Golley F B (ed) *Ecosystems of the World* **14A**. Elsevier, Amsterdam, pp 183–215

Baker H G, Harris B J 1959 Bat-pollination of the silk-cotton tree, *Ceiba pentandra* (L.)Gaertn. (*sensu latu*) in Ghana. *J West African Sci Ass* **5**: 1–9

Barua D N 1969 Seasonal dormancy in tea. *Nature* (*Lond*) **224**: 514

Bawa K S 1982 Patterns of flowering in tropical plants. In Jones C E, Little R J (eds) *Handbook of Pollination Biology*. Van Nostrand/Reinhold, New York

Bazzaz F A, Pickett S T A 1980 Physiological ecology of tropical succession: a comparative review. *Ann Rev Ecol Syst* **11**: 287–310

Becking R W 1957 The Zürich–Montpellier School of phytosociology. *Bot Rev* **23**: 411–88

Bentley B L 1977a The protective function of ants visiting the extrafloral nectaries of *Bixa orellana* (Bixaceae). *J Ecol* **65**: 27–8

Bentley B L 1977b Ectrafloral nectaries and protection by pugnacious bodyguards. *Ann Rev Ecol Syst* **8**: 402–27

Bentley B L 1979 Longevity of individual leaves in a tropical rain forest under-storey. *Ann Bot* **43**: 119–21

Bollinger W H 1980 Sustaining renewable resources: techniques from applied botany. In Campos-Lopéz E (ed) *Renewable Resources: a Systematic Approach*. Academic Press, New York, pp 379–90

Bond T E T 1945 Studies in the vegetative growth and anatomy of the tea plant (*Camellia thea* Link.) with special reference to the phloem. II Further analysis of flushing behaviour. *Ann Bot* **9**: 183–216

Bondad N D, Apostol C J 1979 Induction of flowering and fruiting in immature mango shoots with KNO_3. *Curr Sci* **48**: 591–2

Borchert R 1978 Feedback control and age-related changes of shoot growth in seasonal and non-seasonal climates. In Tomlinson P B, Zimmermann M H (eds) *Tropical Trees as Living Systems*. Cambridge Univ Press, pp 497–516

Borchert R 1983 Phenology and control of flowering in tropical trees. *Biotropica* **15**: 81–9

Bormann F H, Berlyn G (eds) 1981 Age and growth rate of tropical trees: new directions for research. *Yale Univ Sch of Forestry and Env Studies, Bull* **94**: 1–137

Bormann F. H. and Likens G E 1969 The watershed/ecosystem concept and studies of nutrient cycles. In Van Dyne G M (ed) *The Ecosystem Concept in Natural Resource Management*. Academic Press, New York, pp 49–76

Bourlière F 1983 Animal species diversity in tropical forests. In Golley F B (ed) *Tropical Rain Forest Ecosystems*. Elsevier, Amsterdam, pp 77–91

Bowen M R, Howland P, Last F T, Leakey R R B, Longman K A 1977 *Triplochiton scleroxylon*: its conservation and future improvement. *Forest Genet Res Inf*, FAO **6**: 38–47

Brammer H 1962 Soils. In Wills J B (ed) *Agriculture and Land Use in Ghana* Oxford Univ Press, pp 88–126

Bray J R, Gorham E 1964 Litter production in forests of the world. In Cragg J B (ed) *Advances in Ecol Res* **2**: 101–57

Bronchart R 1963 Recherches sur la developpement de *Geophila renaris* de Wild. et Th. Dur. dans les conditions écologiques d'un sous-bois forestier équatorial. Influence sur la mise à fleurs d'une perte en eau disponible du sol. *Mém Soc roy Sci Liège* (*Sér* 5) **8**(2): 1–181

Brown L R 1970 Human food production as a process in the biosphere. *Sci American* Sept 1970: 95–103

Browning G 1977 Environmental control of flower bud development in *Coffea arabica*. In Landsberg J J, Cutting C V (eds) *Environmental Effects on Crop Physiology*. Academic Press, pp 321–31

Browning G 1985 Manipulation of fruit set and the cropping load. In Cannell M G R, Jackson J E (eds) *Attributes of Trees as Crop Plants*. Inst Terrestrial Ecol, Monks Wood, Huntingdon, pp 409–25

Brünig E F 1976 Tree form in relation to environmental conditions: an ecological viewpoint. In Cannell M G R, Last F T (eds) *Tree Physiology and Yield Improvement*. Academic Press, pp 139–56

Brünig E F 1977 The tropical rain forest – a wasted asset or an essential biospheric resource? *Ambio* **6**(4): 187–91

Brünig E F 1983 Vegetation structure and growth in tropical rain forest ecosystems. In Golley F B (ed) *Tropical Rain Forest Ecosystems*. Elsevier, Amsterdam, pp 49–75

Budowski G 1965 Distribution of tropical American rain-forest species in the light of successional processes. *Turrialba* **15**: 40–2

Budowski G 1976 Why save tropical rain forests? Some arguments for campaigning conservationists. *Amazoniana* **4**: 529–38

Budyko M I 1984 *Evolyuciya biosfery*. Gidrometeoizdat, Leningrad

Bullock S H, Bawa K S 1981 Sexual dimorphism and the annual flowering pattern in *Jacaratia dolichaula* (D. Smith)Woodson (Caricaeae) in a Costa Rican rain forest. *Ecology* **62**: 1494–1504

Bünning E 1947 *In den Wäldern Nordsumatras*. F Dümmlers, Bonn

Bunting B T 1967 *The Geography of Soil* (2nd edn). Hutchinson

Burley J, Styles B T (eds) 1975 *Tropical Trees: Variation, Breeding and Conservation*. Academic Press

Burnham C P 1968 Landscapes and soils in Malaya. *Malay Agriculturist* **7**: 64–9

Burtt Davy J 1983 The classification of tropical woody vegetation types. *Imp Forestry Inst, Oxford, Paper* **13**

Cachan P 1963 Signification écolologique des variations microclimatiques verticales dans la forêt sempervirente de Basse Côte d'Ivoire. *Ann Fac Sci, Dakar* **8**: 89–155

Cachan P, Duval J 1963 Variations microclimatiques verticales et saisonnières dans la forêt sempervirente de Basse Côte d'Ivoire. *Ann Fac Sci, Dakar* **8**: 5–87

Cain S A, Oliveria Castro, G M de, Pires J M, da Silva N T 1956 Application of some phytosociological techniques to Brazilian rain forest. *Amer J Bot* **43**: 911–41

Caldwell M M 1981 Plant response to solar ultraviolet radiation. In Lange O L, Nobel P S, Osmund C B, Ziegler H (eds) *Physiological Plant Ecology*. **1**. Springer, Berlin, pp 169–97

Callaham R Z 1962 Geographic variability in growth of forest trees. In Kozlowski T T (ed) *Tree Growth*. Ronald Press, New York, pp 311–25

Campinhos E, Jr, Ikemori Y K 1983 Mass production of *Eucalyptus* spp by rootin(g) cuttings. *Silvicultura* **8**(32): 770–5

Campos-Lopéz E, Neavas-Camacho E 1980 Resource systems evolution. In Campos-Lopéz E (ed) *Renewable Resources: a Systematic Approach*. Academic Press, New York, pp 1–37

Cannell M G R 1979 Biological opportunities for genetic improvement in forest productivity. In Ford E D, Malcolm D C, Atterson J (eds) *The Ecology of Even-Aged Plantations*. Inst Terrestrial Ecol, Cambridge, pp 119–44

Cannell M G R, Njuguna C K, Ford E D, Smith R 1977 Variation in yield among competing individuals within mixed genotype stands of tea: a selection problem. *J Appl Ecol* **14**: 969–86

Carter G S 1934 Reports of the Cambridge expedition to British Guiana, 1933. Illumination in the rain forest at ground level. *J Linn Soc* (*Zool*) **38**: 579–89

Catinot R 1984 En Afrique francophone l'avenir forestier tropical se jouera dans le cadre du monde rural. *Bois et Forêts Tropiques* **203**: 7–43

Chabot B F, Hicks D J 1982 The ecology of leaf life-spans. *Ann Rev Ecol Syst* **13**: 229–59

Chacko E K, Kohli R R, Swamy R D, Randhawa G S 1976 Growth regulators and flowering in juvenile mango (*Mangifera indica* L.). *Acta Horticulturae* **56**: 173–6

Chan H T 1977 Reproductive biology of some Malaysian dipterocarps. *Ph.D. Thesis, Aberdeen Univ*

Chan H T, Appanah S 1980 Reproductive biology of some Malaysian dipterocarps. 1: Flowering biology. *Malaysian Forester* **43**: 132–43

Chang B, Martínez H 1985 Collection of *Gliricidia sepium* seed in central America for provenance trials. *FAO Forest Genet Res Inf* **14**: 50–5

Cherrett J M 1983 Resource conservation by the leaf-cutting ant *Atta cephalotes* in tropical rain forest. In Sutton S L, Whitmore T C, Chadwick A C (eds) *Tropical Rain Forest: Ecology and Management*. Blackwell, pp 253–63

Chipp T F 1927 The Gold Coast forest – a study in synecology. *Oxford Forestry Mem* **7**: 1–94

Chudnoff M 1984 Tropical timbers of the world. *US Dept Agric Handb* **607**

Chunkao K, Tangtham N, Ungkulpakdikul S 1971 Measurements of rainfall in early wet season under hill- and dry-evergreen, natural teak forests of Thailand. *Kog-ma Watershed Res Bull, Bangkok*, no 10

Collins N M 1983 Termite populations and their role in litter removal in Malaysian rain forests. In Sutton S L, Whitmore T C, Chadwick A C (eds) *Tropical Rain Forest: Ecology and Management*. Blackwell, pp 311–16

Collins N M, Morris M G 1985 Threatened swallowtail butterflies of the world. *The IUCN Red Data Book*. International Union for the Conservation of Nature, Gland, Switzerland

Cooling E N G 1968 *Fast Growing Timber Trees of the Lowland Tropics*. **4** Pinus merkusii. Commonw Forestry Inst, Oxford

Coombe D E, Hadfield W 1962 An analysis of the growth of *Musanga cecropioides*. *J Ecol* **50**: 221–34

Corner E J H 1940 Wayside trees of Malaya (2 vols). Govt Printer, Singapore

Corner E J H 1966 *The Natural History of Palms*. Wiedenfeld and Nicolson

Coster C 1923 Lauberneuerung und andere periodische Lebensprozesse in dem trockenen Monsunggebiet Ost-Javas. *Annls Jard bot Buitenz* **33**: 117–89

Coster C 1926 Periodische Blüterscheinungen in der Tropen. *Annls Jard bot Buitenz* **35**: 125–62

Coster C 1927 Die tägliche Schwankungen des Längenzuwachses in den Tropen. Recl Trav bot néer **24**: 257–305

Coster C 1927–28 Zur Anatomie und Physiologie der Zuwachszonen und Jahresringbildung in den Tropen. *Annls Jard bot Buitenz* **37**: 49–160; **38**: 1–114

Coster C 1932 Wortelstudiën in de Tropen. I De jeugdontwikkeling van het wortelstelsel van een zeventigal boomen en groenbemesters. *Tectona* **25**: 828–72

Coster C 1933 Wortelstudiën in de Tropen. III De zuurstof behoefte van het wortelstelsel. *Tectona* **26**: 450–97

Critchfield H J 1966 *General Climatology*. Prentice-Hall, Englewood Cliffs, New Jersey

Croat T B 1975 Physiological behaviour of habit and habitat classes on Barro Colorado Island (Panama Canal Zone). *Biotropica* **7**: 270–7

Damptey H B 1964 *Studies of Apical Dominance in Woody Plants*. MSc Thesis, Univ of Ghana

Darwin C 1859 *On the Origin of Species*. John Murray

Dasilva E J 1980 Trends in microbial technology for developing countries. In Campos-Lopéz E (ed) *Renewable Resources: a Systematic Approach*. Academic Press, New York, pp 329–68

Diamond J M 1973 Distributional ecology of New Guinea birds. *Science* **179**: 759–69

Dick J McP, Longman K A 1985 Techniques for injecting chemicals into trees. *Arboricultural J* **9**: 211–14

Dieterlen F 1978 Zur Phänologie des Äquatorialen Regenwaldes im Ost-Zaire (Kivu), nebst Pflanzenliste und Klimadaten. *Dissertationes Botanicae* **47**, J Cramer, Vaduz, pp 1–111

Doley D 1981 Tropical and subtropical forests and woodlands. In Kozlowski T T (ed) *Water Deficits and Plant Growth. Vol VI Woody Plant Communities.* Academic Press, New York, pp 209–323

Dolph G E, Dilcher D E 1980 Variation in leaf size with respect to climate in Costa Rica. *Biotropica* **12**: 91–9

Doyle T W 1981 The role of disturbance in the gap dynamics of a montane rain forest: an application of a tropical forest succession model. In West D C, Shugart H H, Botkin D B (eds) *Forest Succession, Concepts and Application.* Springer, New York, pp 56–73

Dransfield J 1978 Growth forms of rain forest palms. In Tomlinson P B, Zimmermann M H (eds) *Tropical Trees as Living Systems.* Cambridge Univ Press, pp 247–68

Drosdorff M (ed) 1978 *Diversity of Soils in the Tropics.* ASA/SSSA, Madison, Wisconsin

Dudal R, Moormann F R 1964 Major soils of South-east Asia. *J Trop Geog* **18**: 54–80

Dyanat-Nejad H 1970 Contrôle de le plagiotropie des racines latérales chez *Theobroma cacao* L. *Bull Soc Bot France, Mém* **117**: 183–92

Edens L, Bom I, Ledeboer A M, Maat A, Toonen M Y, Visser C, Verrips C T 1984 Synthesis and processing of the plant protein thaumatin in yeast. *Cell* **37**: 629–33

Edwards P J 1977 Studies of mineral cycling in a montane rain forest in New Guinea. II The production and disappearance of litter. *J Ecol* **65**: 971–92

Edwards P J 1982 V Rates of cycling in throughfall and litter fall. *J Ecol* **70**: 807–27

Ellenberg H 1959a Uber den Wasserhaushalt tropischer Nebeloasen in der Küstenwüste Perus. *Ber Geobot Forsch Inst Rübel, 1958*: 47–74

Ellenberg H 1959b Typen tropischer Urẅalder in Peru. *Schweiz. Z f Forstwesen* **3**: 1–19

Ellenberg H, Mueller-Dombois D 1967 Tentative physiognomic-ecological classification of plant formations of the Earth. *Ber Geobot ForschInst Rübel* **37**: 21–55

Enti A A 1968 Distribution and ecology of *Hildegardia barteri* (Mast.) Kosterm. *Bull de l'IFAN* **30**: 881–95

Ernst W 1971 Zur ökologie der Miombo-Wälder. *Flora* **160**: 317–31

Evans J 1982 *Plantation Forestry in the Tropics.* Clarendon Press

Évrard C 1968 Recherches écologiques sur le peuplement forestier des sols hydromorphes de la cuvette centrale congolaise. *INÉAC, Brussels*, publ **110**: 1–295

Ewel J 1980 Tropical succession: manifold routes to maturity. *Biotropica* **12**: 2–7

Ewel J, Berish C, Brown B, Price N, Raich J 1981 Slash and burn impacts on a Costa Rican wet forest site. *Ecology* **62**: 816–29

Faegri K, van der Pijl L 1979 *The Principles of Pollination Ecology* (3rd edn). Pergamon Press

FAO 1976 *Attempt at a Global Appraisal of the Tropical Moist Forest*. FAO, Rome

FAO 1981 *Tropical Forest Resources Assessment Project (GEMS): Tropical Africa, Tropical Asia, Tropical America* (4 vols) [FAO/UNEP]. FAO, Rome

FAO 1985 *Conservation of Genetic Resources: Data Book on Endangered Arboreal Species and Provenances*. FAO, Rome

FAO/UNESCO 1968 Definitions of soil units for the Soil Map of the World. *World Soil Resources Rep* **33**. [also suppl of 24 Sept 1970] FAO, Rome

Fernandez I R, Ocampo A T 1980 Considerations on the social impact of technology. In Campos-Lopéz E (ed) *Renewable Resources: a Systematic Approach*. Academic Press, New York, pp 223–32

Fetcher N, Oberbauer S F, Strain B R 1985 Vegetation effects on microclimate in lowland tropical forest in Costa Rica. *Internat J Biometeor* **29**: 145–55

Fink S 1982 Histochemische Untersuchungen über Stärkeverteilung und Phosphataseaktivität im Holz einiger tropischer Baumarten. *Holzforshung* **36**: 295–302

Fisher J B 1978 A quantitative study of *Terminalia* branching. In Tomlinson P B, Zimmermann M H (eds) *Tropical Trees as Living Systems*. Cambridge, pp 285–320

Fisher J B 1984 Tree architecture: relationships between structure and function. In White R A, Dickison W C (eds) *Contemporary Problems in Plant Anatomy*. Academic Press, Orlando, Florida, pp 541–89

Flenley J R 1979 *The Equatorial Rain Forest: a Geological History*. Butterworths

Flores Rodas M A 1983 *Food and Fruit-Bearing Forest Species. 1 Examples from Eastern Africa*. FAO, Forestry Paper **44/1**, Rome

Ford E D 1976 Competition, genetic systems and improvement of forest yield. In Cannell M G R, Last F T (eds) *Tree Physiology and Yield Improvement*. Academic Press, pp 463–72

Fosberg F R 1973 Temperate zone influence on tropical forest land use. In Meggers B J, Ayensu E S, Duckworth W D *Tropical Forest Ecosystems in Africa and South America: a*

Comparative Review Smithsonian Inst, Washington, DC, pp 345–50

Foster R B 1977 *Tachigalia versicolor* is a suicidal neotropical tree. *Nature (Lond)* **268**: 624–6

Foster R B 1982a The seasonal rhythm of fruitfall on Barro Colorado Island.

Foster R B 1982b Famine on Barro Colorado Island. In Leigh E G, Jr, Rand A S, Windsor D M (eds) *The Ecology of a Tropical Forest – Seasonal Rhythms and Long-Term Changes.* Smithsonian Inst, Washington, DC, pp 151–72; 201–11

Foxworthy F W 1927 Commercial timber trees of the Malay Peninsula. *Malay Forest Rec* **3**

Frankel O H, Bennett E (eds) 1970 *Genetic Resources in Plants – Their Exploration and Conservation.* (IBP Handb **11**) Blackwell

Frankie G. W, Baker H G, Opler P A 1974 Comparative phenological studies of trees in tropical wet and dry forests in the lowlands of Costa Rica. *J Ecol* **62**: 881–919

Gaillochet J, Gavaudan P 1974 Comportement nycthemeral des feuilles de *Cassia fasciculata* Michx.: influence de la temperature et de la glande petiolaire. *C R Séances Soc Biol Fil* **168**: 1094–1101

Gentry A H 1974 Flowering phenology and diversity in tropical Bignoniaceae. *Biotropica* **6**: 64–8

Germain R 1965 Les biotopes alluvionnaires herbeux et les savanes intercalaires du Congo équatorial. *Acad Roy Sci d'Outre-Mer, Cl Sci nat et Médic* NS **15**(4): 1–399

Germain R, Évrard C 1956 Étude écologique et phytosociologique de la forêt à *Brachystegia laurentii. INÉAC, Brussels, publ* **67**: 1–105

Gill A M 1971 Endogenous control of growth-ring development in *Avicennia. Forest Sci* **17**: 462–5

Gill A M, Tomlinson P B 1971 Studies on the growth of red mangrove (*Rhizophora mangle* L.). 3 Phenology of the shoot. *Biotropica* **3**: 109–24

Gill A M, Tomlinson P B 1975 Aerial roots: an array of forms and functions. In Clarkson D T, Torrey J G (eds) *The Development and Function of Roots.* Academic Press, pp 237–60

Gillett J B 1962 Pest pressure, an underestimated factor in evolution. *Publs Syst Ass, London* **4**: 37–46

Goldstein I S 1980 The conversion of renewable resources into chemical feedstocks. In Campos-Lopéz E (ed) *Renewable Resources: a Systematic Approach.* Academic Press, New York, pp 275–86

Golley F B (ed) 1983a *Ecosystems of the World.* **14A** *Tropical rain forest ecosystems,* Elsevier, Amsterdam

Golley F B 1983b The abundance of energy and chemical elements. In Golley F B (ed) 1983a *Tropical Rain Forest Ecosystems*. Elsevier, Amsterdam, pp 101–15

Golley F B, McGinnis J T, Clements R G, Child G I, Duever M J 1975 *Mineral Cycling in a Tropical Moist Forest Ecosystem*. Univ Georgia Press, USA

Gómez-Pompa A 1980 Renewable resources from the tropics. In Campos-Lopéz E (ed) *Renewable Resources: a Systematic Approach*. Academic Press, New York, pp 391–406

Gómez-Pompa A, Vázquez-Yanes C, del Amo Rodríguez S, Butanda Cervera A (eds) 1976 *Investigaciones sobre la regeneración de selvas altas en Veracruz, Mexico*. Compañia Editorial Continental S A, Mexico

Goodland R J 1975 The tropical origin of ecology: Eugen Warming's jubilee. *Oikos* **26**: 240–5

Greathouse D C, Laetsch W M 1973 Effects of gibberellin A_3 on the dimorphic branch system of *Theobroma cacao, Plant Sci Lett* **1**: 39–45

Greenland D J (ed) 1981 *Characterization of Soils in Relation to their Classification and Management for Crop Production: Examples from the Humid Tropics*. Oxford Univ Press

Greenwood M, Posnette A F 1949 The growth flushes of cocoa. *J Hort Sci* **25**: 164–74

Greenwood M S 1981 Reproductive development in loblolly pine. II The effect of age, gibberellin plus water stress and out-of-phase dormancy on long shoot growth behavior. *Amer J Bot* **68**: 1184–90

Grison F 1978 Research into the floral biology of *Aucoumea klaineana* Pierre. In Nikles D G, Burley J, Barnes R D (eds) *Progress and Problems of Genetic Improvement of Tropical Forest Trees*. Commonw Forestry Inst, Oxford, pp 921–6

Grout W W, Shelton K, Pritchard H W 1983 Orthodox behaviour of oil palm seed and cryopreservation of the excised embryo for genetic conservation. *Ann Bot* **52**: 381–4

Grubb P J 1977 Control of forest growth and distribution on wet tropical mountains: with special references to mineral nutrition. A *Rev Ecol Syst* **8**: 83–107

Guardiola J L, Monerri C, Agusti, M 1982 The inhibitory effect of gibberellic acid on flowering in citrus. *Physiol plantarum* **55**: 136–42

Guillaumet J L 1967 *Recherches sur la végétation et la flore de la région du Bas-Cavally (Côte d'Ivoire)*. ORSTOM, Paris

Gutiérrez M 1978 Application of gibberellin to promote and increase the flowering of *Cupressus lusitanica. Cellulosa y*

Papel de Colombia S A, Cali, Investigación Forestal, Res Rep **39**: 1–6

Hackett W P 1985 Juvenility, maturation and rejuvenation in woody plants. *Hortic Rev* **7**: 109–55

Haffer J J 1982 General aspects of the refuge theory. In Prance G T (ed) *Biological Diversification in the Tropics*. Columbia Univ Press, New York, pp 6–24

Hall J B, Swaine M D 1976 Classification and ecology of close-canopy forest in Ghana. *J Ecol* **64**: 913–51

Hall J B, Swaine M D 1980 Seed stocks in Ghanaian forest soils. *Biotropica* **12**: 256–63

Hall J B, Swaine M D 1981 *Distribution and ecology of vascular plants in a tropical rain forest. Forest vegetation in Ghana*. Junk, The Hague

Hall J B, Okali D U U, Lock J M 1975 Night illumination and flowering in an indigenous West African forest-tree, *Hildegardia barteri* (Mast.) Kosterm. *Bull de l'IFAN* **37**(A,2): 282–7

Hallé F, Martin R 1968 Étude de la croissance rythmique chez l'Hévéa *Hevea brasiliensis* (Müll. Arg. Euphorbiacées-Crotonoidées). *Adansonia* (NS) **8**: 475–503

Hallé F, Oldeman R A A 1970 *Essai sur l'architecture et la dynamique de croissance des arbres tropicaux*. Masson, Paris (Engl transl by Stone B C 1975. Penerbit Univ Malaysia, Kuala Lumpur)

Hallé F, Oldeman R A A, Tomlinson P B 1978 *Tropical Trees and Forests: an Architectural Analysis*. Springer, Berlin

Hardwick K, Abo-Hamed S, Collin H A 1982 Hormonal control of shoot apex activity in *Theobroma cacao L. Proc 8th Internat Cocoa Res Conf, Cartagena 1981*, pp 259–64

Hardwick K, Collin H A 1985 Plant water status and the control of leaf production in cocoa. *Proc 9th Internat Cocoa Res Conf, Lomé 1984*, pp 111–16

Hardwick K, Machado R C R, Smith J, Veltkamp C J 1985 Apical bud activity in cocoa. *Proc 9th Internat Cocoa Res Conf, Lomé 1984*, pp 153–58

Harrington J F 1970 Seed and pollen storage for conservation of plant gene resources. In Frankel O H, Bennett E (eds) *Genetic Resources in Plants – Their Exploration and Conservation*. Blackwell, pp 501–21

Hedberg I *et al* 1982–83 Inventory of plants used in traditional medicine in Tanzania. *J Ethnopharmacology* **6**: 29–60; **9**: 105–28; **9**: 237–60

Hedegart T 1976 Breeding systems, variation and genetic improvement of teak (*Tectona grandis* L.f.). In Burley J, Styles B T (eds) *Tropical Trees: Variation, Breeding and Conservation.*

Academic Press, 109–23
Heinsdijk D, Miranda Bastos A de 1963 Inventários florestais na Amazônia. *Bol Setor Inventários Florestais (Brazília)* **6**: 1–100
Heybroek H M 1978 Primary considerations: multiplication and genetic diversity. *Unasylva* **30**: 27–33; 49–50
Holdsworth M 1961 The flowering of rain flowers. *J W Afr Sci Assoc* **7**: 28–36
Holttum R E 1953 Evolutionary trends in an equatorial climate. *Symp Soc exptl Biol* **7**: 159–73
Hopkins B 1970 Vegetation of the Olokemeji Forest Reserve, Nigeria. VI. The plants on the forest site, with special reference to their seasonal growth. *J Ecol* **58**: 765–93
Houaye K P 1983 Quelques essais de multiplication végétative de jeunes plantes de teck élevés en conditions artificielles. *Diplôme, Univ Nancy, France*
Howland P 1975 Current management techniques for raising *Triplochiton scleroxylon* K Schum from stem cuttings. In Proc Symp *Variation and Breeding Systems of* Triplochiton scleroxylon *K Schum*, Forest Res Inst Nigeria, Ibadan, pp 125–9
Howland P, Bowen M R 1977 *West African Hardwoods Improvement*, Project Res Rep 1971–77, Forest Res Inst Nigeria, Ibadan
Hubbell S 1979 Tree dispersion, abundance, and diversity in a tropical dry forest. *Science* **203**: 79–89
Hubbell S P, Foster R B 1983 Diversity of canopy trees in a neotropical forest and implications for conservation. In Sutton S L, Whitmore T C, Chadwick A C (eds) *Tropical Rain Forest: Ecology and Management*. Blackwell, pp 25–41
Hueck K 1966 *Die Wälder Südamerikas*. G Fischer, Stuttgart
Hutchinson I D 1982 Approach adopted to define interim guidelines for silviculture and management of mixed dipterocarp forest in Sarawak. In Srivastava P B L *et al.* (eds) *Tropical Forests – Source of Energy through Optimisation and Diversification*. Univ Pertanian Malaysia, Serdang, pp 127–46
Hutchinson J, Dalziel J M, Keay R W J 1954–72 *Flora of West Tropical Africa* (2nd edn.) **1**(1) 1954; **1**(2) 1958; **2** 1963; **3**(1) 1968; **3**(2) 1972. Crown Agents
Huxley P A (ed) 1983 *Plant Research and Agroforestry*. ICRAF, Nairobi
IITA 1978 *Research Highlights 1977*, Internat Inst Trop Agric, Ibadan
IITA 1980 *Research Highlights 1979*, Internat Inst Trop Agric, Ibadan
Irvine F R 1961 *Woody Plants of Ghana*. Oxford Univ Press
Ison, R L, Humphries L R 1984 Flowering of *Stylosanthes*

guianensis in relation to juvenility and the long-short day requirement. *J Exp Bot* **35**: 121–6

Ivory M H, Munga F M 1983 Growth and survival of container-grown *Pinus caribaea* infected with various ectomycorrhizal fungi. In Atkinson D, Bhat K K S, Coutts M P, Mason P A, Read D J (eds) *Tree Root Systems and Their Mycorrhizas*. Junk, The Hague

Janos D O 1983 Tropical mycorrhizas, nutrient cycles and plant growth. In Sutton S L, Whitmore T C, Chadwick A C (eds) *Tropical Rain Forest: Ecology and Management*. Blackwell, pp 327–45

Janzen D H 1967 Synchronisation of sexual reproduction of trees within the dry season in Central America. *Evolution* **21**: 620–37

Janzen D H 1973 Tropical agroecosystems. *Science* **182**: 1212–19

Janzen D H 1974 Tropical blackwater rivers, animals and mast fruiting by the Dipterocarpaceae. *Biotropica* **6**: 69–103

Janzen D H 1975 *Ecology of Plants in the Tropics*. Edward Arnold

Janzen D H 1976a The microclimate difference between a deciduous forest and adjacent riparian forest in Guanacaste province, Costa Rica. *Brenesia* **8**: 29–33

Janzen D H 1976b Why are there so many species of insects? *Proc Internat Congr Entomol 1976*, pp 84–94

Janzen D H 1976c Why bamboos wait so long to flower. *Ann Rev Ecol Syst* **7**: 347–91

Janzen D H 1982 Ecological distribution of chlorophyllous developing embryos among perennial plants in a tropical deciduous forest. *Biotropica* **14**: 232–6

Janzen D H, Wilson D E 1974 The cost of being dormant in the tropics. *Biotropica* **6**: 260–2

Jeník J 1969 The life-form of *Scaphopetalum amoenum* A. Chev. *Preslia* **41**: 109–12

Jeník J 1971 Root system of tropical trees. 7 The facultative peg-roots of *Anthocleista nobilis* G. Don. *Preslia* **43**: 97–104

Jeník J 1973 Root system of tropical trees. 8 Stilt-roots and allied adaptations. *Preslia* **45**: 250–64

Jeník J 1978 Roots and root systems in tropical trees: morphologic and ecologic aspects. In Tomlinson P B, Zimmermann M H (eds) *Tropical Trees as Living Systems*. Cambridge Univ Press, pp 332–49

Jeník J, Enti A A 1969 Discontinuous distribution of *Allexis cauliflora* (Oliv.) L. Pierre in equatorial Africa. *Novit Bot, Inst Bot Univ Carol Prag, 1968*: 67–71

Jeník J, Hall J B 1966 The ecological effects of the harmattan wind in Djebobo Massif, Togo. *J Ecol* **54**: 767–79

Jeník J, Harris B J 1969 Root-spines and spine-roots in dicotyledonous trees of tropical Africa. *Öst Bot Z* **117**: 128–38

Jones E W 1955–56 Ecological studies on the rain forest of southern Nigeria. IV The plateau forest of the Okumu Forest Reserve (Pts I & II). *J Ecol* **43**: 564–94; **44**: 83–117

Jordan C F (ed) 1981 *Tropical Ecology*, Hutchinson Ross, Stroudsburg PA, p 244

Jordan C F 1983 Productivity of tropical rain forest ecosystems and the implications for their use as future wood and energy sources. In Golley F B (ed) *Ecosystems of the World.* **14A** *Tropical Rain Forest Ecosystems.* Elsevier, Amsterdam, pp 117–36

Jordan C F 1985 *Nutrient cycling in tropical forest ecosystems.* Wiley.

Jordan E L 1981 The birth of ecology: an account of Alexander von Humboldt's voyage to the equatorial regions of the new continent. In Jordan C F (ed) *Tropical Ecology*, Hutchinson Ross, Stroudsburg PA, pp 4–15

Jose A I, Koshy M M 1972 A study of the physical and chemical characteristics of soils as influenced by teak vegetation. *Indian Forester* **98**: 338–48

Kaur A *et al* 1978 Apomixis may be widespread among trees of the climax rain forest. *Nature (Lond)* **271**: 440–2

Kellmann M, Hudson J, Sanmugas K 1982 Temporal variability in atmospheric nutrient influx to a tropical ecosystem. *Biotropica* **14**: 1–9

Kelly P M, Sear C B 1984 Climatic impact of explosive volcanic eruptions. *Nature (Lond)* **311**: 740–3

Keong T C 1983 *Acacia mangium* Willd. – a plantation species for *Imperata cylindrica* (L.)Beauv. grassland in Sabah. *Silvicultura* **8**(30): 321–6

King K F S 1979 Agroforestry and the utilisation of fragile ecosystems. *Forest Ecol Manage* **2**: 161–8

King M W, Roberts E H 1979 *The Storage of Recalcitrant Seeds – Achievements and Possible Approaches.* Internat Board for Plant Genet Res, FAO, Rome

Kira T, Shidei T 1967 Primary production and turnover of organic matter in different forest ecosystems of the western Pacific. *Jap J Ecol* **17**: 70–87

Kitazawa Y 1971 Biological regionality of the soil fauna and its function in forest ecosystem types. In Duvigneaud P (ed) *Productivity in Forest Ecosystems*. UNESCO, Paris, pp 485–98

Klebs G 1926 Über periodisch wachsende tropische Baumarten. *Sber heidelb Akad Wiss Math-Naturwiss Kl* **2**: 1–31

Klinge H 1973 Struktur und Artereichtum des Zentralamazonischen Regenwaldes. *Amazonia* **4**: 283–92

Köppen W 1930 *Klimate der Erde*. de Gruyter, Berlin

Koptur S, Smith A R, Baker I 1982 Nectaries in some neotropical species of Polypodium (Polypodiaceae). Preliminary observations and analyses. *Biotropica* **14**: 108–13

Koranski D S, McCown B H, Struckmeyer B E, Beck G E 1979 Gibberellin-growth retardant interactions on the growth and flowering of Clerodendrum thompsoniae. *Physiol Plant* **45**: 88–92

Koriba K 1958 On the periodicity of tree-growth in the tropics, with reference to the mode of branching, the leaf-fall, and the formation of the resting bud. *Gdns' Bull, Singapore* **17**: 11–81

Kramer P J, Kozlowski T T 1960 *Physiology of Trees* (1st edn). McGraw Hill, New York

Kuijt J 1969 *The Biology of Parasitic Flowering Plants*. Univ California Press, Berkeley

Kursanov A L 1960 The physiology of the whole plant. *Wye Coll, Univ London, Occasional Publ* **12**: 1–20

Kwakwa R S 1964 *The Effects of Temperature and Day-Length on Growth and Flowering in Woody Plants*. MSc Thesis, Univ of Ghana

Kwesiga F R 1985 *Aspects of the Growth and Physiology of Tropical Tree Seedlings in Shade*. PhD thesis, Univ of Edinburgh

Ladipo D O, Grace J, Sandford A P, Leakey R R B 1984 Clonal variation in photosynthetic and respiration rates and diffusion resistances in the tropical hardwood *Triplochiton scleroxylon* K. Schum. *Photosynthetica* **18**: 20–27

Ladipo D O, Leakey R R B, Last F T 1983 A study of variation in *Triplochiton scleroxylon* K. Schum. – some criteria for clonal selection. *Silvicultura* **8**(30): 333–6

Lal R 1981 Management of soils for continuous production: controlling erosion and maintaining physical condition. In Greenland D J (ed) *Characterisation of Soils*. Oxford Univ Press, pp 188–201

Lamont B 1981 Availability of water and inorganic nutrients in the persistent leaf bases of the grasstree *Kingia australis*, and the uptake and translocation of labelled phosphate by the embedded aerial roots. *Physiol Plant* **52**: 181–6

Lang G E, Knight D H 1979 Decay rates for boles of tropical trees in Panama. *Biotropica* **11**: 316–17

Larcher W 1980 *Ecological Plant Physiology* (2nd edn). Springer, Berlin

Lawson G W, Armstrong-Mensah K O, Hall J B 1970 A catena in moist semi-deciduous forest near Kade, Ghana. *J Ecol* **58**: 371–98

Leakey L S B 1964 Prehistoric man in the tropical environment. In *The Ecology of Man in the Tropical Environment* IUCN, Morges, pp 24–9

Leakey R R B 1983 Stockplant factors affecting root initiation in cuttings of *Triplochiton scleroxylon* K. Schum., an indigenous hardwood of West Africa. *J Hort Sci* **58**: 277–90

Leakey R R B 1985a The capacity for vegetative propagation in trees. In Cannell M G R, Jackson J E (eds) *Attributes of Trees as Crop Plants*. Inst Terrestrial Ecol, Monks Wood, Huntingdon, pp 110–33

Leakey R R B 1985b Prediction of branching habit in clonal *Triplochiton scleroxylon*. In Tigerstedt P M A, Puttonen P, Koski V (eds) *Crop Physiology of Forest Trees*. Univ of Helsinki, Viiki, Finland

Leakey R R B, Ferguson N R, Longman K A 1981 Precocious flowering and reproductive biology of *Triplochiton scleroxylon* K. Schum. *Commw Forestry Rev* **60**: 117–26

Leakey R R B, Chapman V R, Longman K A 1982a Physiological studies for tropical tree improvement and conservation. Some factors affecting root initiation of cuttings of *Triplochiton scleroxylon* K. Schum. *For Ecol Manage* **4**: 53–66

Leakey R R B, Grison F 1985 *Regional Programme for the Improvement of Tropical Hardwoods – West and Central Africa*. Prep Phase Rep to European Development Fund, Inst Terrestrial Ecol, Penicuik, pp 1–209

Leakey R R B, Last F T, Longman K A 1982b Domestication of tropical trees: an approach securing future productivity and diversity in managed ecosystems. *Commonw Forestry Rev* **61**: 31–42

Leakey R R B, Mohammed H R S 1985 The effects of stem length on root initiation in sequential single-node cuttings of *Triplochiton scleroxylon* K. Schum. *J Hort Sci* **60**: 431–7

Lebrun J 1947 La végétation de la plaine alluviale au sud du Lac Édouard. (Expl Parc Nat Albert, mission J Lebrun, 1937–38.) *Inst Parcs Nat Congo Belge* **1**: 1–800

Lebrun J, Gilbert G 1954 Une classification écologique des forêts du Congo. *INÉAC, Brussels, Publ* **63**: 1–89

Lee D H, Lowry J B 1974 Physical basis and ecological significance of iridescence in blue plants. *Nature* (Lond) **254**: 50–1

Lee D W, Lowry J B, Stone B C 1979 Abaxial anthocyanin layer in leaves of tropical rain forest plants: enhancer of light capture in deep shade. *Biotropica* **11**: 70–7

Lees H S 1980 *Shorea pinanga* Burck. flowers 6 years after planting in research plot no. 76. *Malaysian Forester* **43**: 126–7

Leigh E G, Jr 1982 Introduction: why are there so many kinds of tropical trees. In Leigh E G, Rand A S, Windsor D M (eds) *The Ecology of a Tropical Forest – Seasonal Rhythms and Long-Term Changes*. Smithsonian Inst, Washington, DC, pp 63–6

Leigh E G, Jr, Windsor D M 1982 Forest production and regulation of primary consumers on Barro Colorado Island. In Leigh E G, Rand A S, Windsor D M (eds) *The Ecology of a Tropical Forest – Seasonal Rhythms and Long-term Changes*. Smithsonian Inst, Washington, DC, pp 111–22

Lemée G 1956 Recherches écophysiologiques sur les cacaoyers. *Revue Gén Bot* **63**: 41–94

Leroy Deval J 1974 *Structure dynamique de la rhizosphere de l'okoumé dans ses rapports avec la sylviculture*. Centre Tech Forestier Trop, Nogent-sur-Marne

Libby W J 1982 What is a safe number of clones per plantation? In *Resistance to Pests and Diseases in Forest Trees*. Proc 3rd internat wkshp on genet of host/parasite interactions in forest trees, Wageningen, Netherlands, Sept 1980

Libby W J 1985 Potential of clonal forestry. In Zsuffa L, Rauter R M, Yeatman C W (eds) *Clonal Forestry: Its Impact in Tree Improvement and Our Future Forests*. Canad Forest Serv, Petawawa, Ontario, pp 1–11

Lieberman D, Hall J B, Swaine M D, Lieberman M 1979 Seed dispersal by baboons in the Shai Hills, Ghana. **60**: 65–75

Lieberman D, Lieberman M, Hartshorn G, Peralta R 1985 Growth rates and age–size relationships of tropical wet forest trees in Costa Rica. *J Trop Ecol*. **1**: 97–109

Lieth H 1970 Phenology in productivity studies. In Reichle D (ed) *Analysis of Temperate Forest Ecosystems*. Springer, Heidelberg, pp 29–46

Lieth H 1974 *Phenology and Seasonality Modelling*. Springer, Berlin

Livingstone D A, van der Hammen T 1978 Paleogeography and paleoclimatology. In *Tropical Forest Ecosystems*. UNESCO, Paris, pp 61–90

Lo Y-N 1985 Root initiation of *Shorea macrophylla* cuttings: effects of node position, growth regulators and misting regime. *Forest Ecol Manage* **12**: 43–52

Löhr E, Müller D 1968 Blatt-Atmung der höheren Bodenpflanzen in tropischen Regenurwald. *Physiol Plant* **21**: 673–5

Longman K A 1969 The dormancy and survival of plants in the humid tropics. *Symp Soc Exp Biol* **23**: 471–88

Longman K A 1978 Control of shoot extension and dormancy –

external and internal factors. In Tomlinson P B, Zimmermann M H (eds) *Tropical Trees as Living Systems*. Cambridge Univ Press, New York, pp 465–95

Longman K A 1982 Ecological and physiological foundations for managing tropical forest resources. In Srivastava P B L *et al.* (eds) *Tropical Forests – Source of Energy through Optimisation and Diversification*. Univ Pertanian Malaysia, Serdang, pp 63–75

Longman K A 1985 Tropical forest trees. In Halevy A H (ed) *Handbook of Flowering* (Vol 1). CRC Press, Boca Raton, Fla, pp 23–39

Longman K A 1986 The significance of juvenility for seed orchard management. *Forest Ecol Manage*. (in press)

Longman K A, Coutts M P 1974 Physiology of the oak tree. In Morris M G, Perring F H (eds) *The British Oak*. BSBI/Classey, pp 194–221

Longman K A, Leakey R R B, Howland P, Bowen M R 1978 Physiological approaches for utilizing and conserving the genetic resources of tropical trees. *Proc 3rd World Consult Forest Tree Breeding* (Vol 2). CSIRO, Canberra, pp 1043–54

Longman K A, Leakey R R B, Denne M P 1979 Genetic and environmental effects on shoot growth and xylem formation in a tropical tree. *Ann Bot* **44**: 377–80

Longman K A, Dick J McP, Page C N 1982 Cone induction with gibberellin for taxonomic studies in Cupressaceae and Taxodiaceae. *Biol Plant* **24**: 195–201

Luckwill L C 1981 *Growth Regulators in Crop Production*. Edward Arnold

Lyr H, Hoffmann G 1967 Growth rates and growth periodicity of tree roots. *Internat Rev Forest Res* **2**: 181–236

Lyons J M 1973 Chilling injury in plants. *Ann Rev Plant Physiol* **24**: 445–66

Mabberley D J 1983 *Tropical Rain Forest Ecology*. Blackie

MacKinnon J A 1972 *The Behaviour and Ecology of the Orang Utan* (Pongo pygmaeus). D Phil Thesis, Oxford University

Madge D S 1965 Leaf fall and litter disappearance in a tropical forest. *Pedobiologia* **5**: 273–88

Madison M 1977 Vascular epiphytes: their systematic occurrence and salient features. *Selkyana* **2**: 1–13

Malingreau J P, Stephens G, Fellows L 1985 Remote sensing of forest fires: Kalimantan and North Borneo in 1982–83. *Ambio* **14**: 314–321

Manokaran N 1979 Stemflow, throughfall and rainfall interception in a lowland tropical rain forest in Peninsular Malaysia. *Malays Forester* **42**: 174–201

Mariaux A 1981 Past efforts at measuring age and annual growth in tropical trees. In Bormann F H, Berlyn G (eds) *Age and Growth Rate of Tropical Trees: New Directions for Research.* Yale Sch For & Env Studies, Bull **94**: 20–30

Markham R H, Babbedge A J 1979 Soil and vegetation catenas on the forest–savanna boundary in Ghana. *Biotropica* **11**: 224–34

Marn H M, Jonkers W 1982 Logging damage in tropical high forest. In Srivastava *et al.* (eds) *Tropical Forests: Source of Energy through Optimisation and Diversification.* Univ Pertanian Malaysia, Serdang, pp 27–38

Marzola D L, Bartholomew D P 1979 Photosynthetic pathway and biomass energy production. *Science* **205**: 555–9

Mascarenhas *et al*, 1982 Propagation of trees by tissue culture. In Rao A N (ed) *Tissue Culture of Economically Important Plants.* Cttee Sci Tech Dev Countries/Asian Network for Biol Sci, Singapore, pp 175–9

Mathon C C 1975 *Écologie du développement et phytogéographie.* Faculté des Sciences de l'Université de Poitiers, France

Mayr E 1982 *The Growth of Biological Thought: Diversity, Evolution and Inheritance.* Harvard Univ Press, Cambridge, Mass

Mazalewski R L, Hackett, W P 1979 Cutting propagation of *Eucalyptus ficifolia* using cytokinin-induced basal, trunk shoots. *Proc Internat Plant Propagators Soc* **29**: 118–25

Medina E 1983 Adaptations of tropical trees to moisture stress. In Golley F B (ed) *Ecosystems of the World.* **14A** *Tropical Rain Forest Ecosystems.* Elsevier, Amsterdam, pp 225–37

Medina E, Klinge H 1983 Productivity of tropical forests and tropical woodlands. *Encycl Plant Physiol* (NS) **12D**: 281–303

Medina E, Mooney H, Vázquez-Yanes C (eds) 1984 *Physiological Ecology of Plants of the Wet Tropics.* Junk, The Hague

Medway Lord 1972 Phenology of a tropical rain forest in Malaya. *Biol J Linn Soc* **4**: 117–46

Meentemeyer V 1978 An approach to the biometeorology of decomposer organisms. *Internat J Biometeorol* **22**: 94–102

Meggers B J, Ayensu E S, Duckworth W D (eds) 1973 *Tropical Forest Ecosystems in Africa and South America: a Comparative Review*, Smithsonian Inst, Washington, DC

Melville R 1970 Plant Conservation and the Red Data Book. *Biol Cons* **2**: 185–8

Mensah K O A, Jeník J 1968 Root system of tropical trees. 2 Features of the root system of iroko (*Chlorophora excelsa* Benth. et Hook). *Preslia* **40**: 21–7

Michon G 1983 Village – forest – gardens in west Java. In Huxley

P A (ed) *Plant Research and Agroforestry*. Internat Council for Res in Agrofor, Nairobi, pp 13–24

Miller D B, Feddes R G 1971 *Global Atlas of Relative Cloud Cover 1967–70*. Dept of Commerce, NOAA, Washington, DC

Miller L D, Nualchawee K, Tom C 1979 Shifting cultivation in the forests of northern Thailand. In Williams D L, Miller L D (eds) *Monitoring Forest Canopy Alteration around the World with Digital Analysis of Landsat Imagery*. NASA, Washington, DC, pp 35–45

Mitchell A W 1982 *Reaching the Rain Forest Roof*. Leeds Phil & Lit Soc

Monteith J L 1972 Solar radiation and periodicity in tropical ecosystems. *Ann Appl Ecol* **9**: 747–66

Moorman F R, Kang B T 1978 Microvariability of soils in the tropics and its agronomic implications with special reference to West Africa, In Drosdorff M (ed) *Diversity of Soils in the Tropics*. ASA/SSSA, Madison, Wisconsin, pp 29–43

Morgan D C, Smith H 1981 Non-photosynthetic responses to light quality. *Encycl Plant Physiol* (NS) **12A**: 108–34

Morrison C G T, Hoyle A C, Hope-Simpson J F 1948 Tropical soil-vegetation catenas and mosaics. *J Ecol* **36**: 1–84

Murashige T 1966 The deciduous behaviour of a tropical plant, *Plumeria acuminata. Physiol Plant* **19**: 348–55

Murphy P G 1975 Net primary productivity in tropical terrestrial ecosystems. In Lieth H, Whittaker R H (eds) *Primary Productivity of the Biosphere*. Springer, New York, pp 217–31

Murray D B 1966 Soil moisture regimes. *Cacao Res, Trinidad, Ann Rep 1965*, pp 34–9

Murray D B, Sale P J M 1966 Report on plant physiology. *Cacao Res, Trinidad, Ann Rep 1965*, pp 30–4

Murray D B, Sale P J M 1967 Growth studies on cacao in controlled environment rooms. Conf internat recherches agronomiques cacaoyeres, Abidjan 1965, pp 57–63. Paris

Myers N 1983a Conversion rates in tropical moist forests. In Golley F B (ed) *Ecosystems of the World*. **14A** *Tropical Rain Forest Ecosystems*. Elsevier, Amsterdam, pp 289–300

Myers N 1983b Conservation of rain forests for scientific research, for wildlife conservation, and for recreation and tourism. In Golley F B (ed) *Tropical Rain Forest Ecosystems*. Elsevier, Amsterdam, pp 325–34

Myers R L 1983c Site susceptibility to invasion by the exotic tree *Melaleuca quiquenervia* in southern Florida. *J Appl Ecol* **20**: 645–58

Nagarajah S, Ratnasuriya G B 1981 Clonal variability in root growth and drought resistance in tea (*Camellia sinensis*).

Plant and Soil **60**: 153–5
NAS 1975 *Underexploited Tropical Plants with Promising Economic Value*. BOSTID publ **16**, Nat Acad Sci, Washington, DC
NAS 1979 *Tropical legumes: resources for the future*. BOSTID publ **25**, Nat Acad Sci, Washington, DC
NAS 1980 *Firewood Crops: Shrub and Tree species for Energy Production*. BOSTID publ **27**, Nat Acad Sci, Washington, DC
NAS 1982 *Ecological Aspects of Development in the Humid Tropics*. Nat Acad Sci, Washington, DC
NAS 1983a *Firewood crops: Shrub and Tree Species for Energy Production* Vol 2. BOSTID publ **40**, Nat Acad Sci, Washington, DC
NAS 1983b *Casuarinas: Nitrogen-Fixing Trees for Adverse Sites*. BOSTID publ **43**, Nat Acad Sci, Washington, DC
NAS 1984 *Leucaena: Promising Forage and Tree Crop in Developing Countries*. BOSTID publ **52**, Nat Acad Sci, Washington, DC
Nations J D, Nigh R B 1980 The evolutionary potential of Lacandon Maya sustained-yield tropical forest agriculture. *J Anthropol Res* **36**: 1–30
Naundorf G 1954 Contribution a la fisiologia de la floracion en cacao. Existencia de hormonas de floracion. *Cacao en Colombia* **3**: 29–34
Ng F S P 1966 Age at first flowering of dipterocarps. *Malay Forester* **29**: 290–5
Ng F S P 1975 A note on natural root grafts in Malaysian trees. *Malays Forester* **38**: 153–9
Ng F S P 1978 Strategies for establishment in Malayan forest trees. In Tomlinson P B, Zimmermann M H (eds) *Tropical Trees as Living Systems*. Cambridge Univ Press, pp 129–62
Ng F S P, Loh H S 1975 Flowering-to-fruiting periods *Malays Forester* **37**: 127–32
Nicholson D L 1965 A plot of *Albizzia falcata* (L.)Backer. *Malay Forester* **28**: 240–2
Nishizawa N, Mori S 1977 Invagination of plasmalemma, its role in the absorption of macromolecules in rice roots. *Plant and Cell Physiol* **18**: 767–82
Njoku E 1963 Seasonal periodicity in the growth and development of some forest trees in Nigeria. *J Ecol* **51**: 617–24
Njoku E 1964 Seasonal periodicity in the growth and development of some forest trees in Nigeria. 2 Observations on seedlings. *J Ecol* **52**: 19–26
Nye P H 1961 Organic matter and nutrient cycles under moist

tropical forest. *Plant and Soil* **13**: 333–46
Nye P H, Greenland D J 1960 *The Soil under Shifting Cultivation*. Commonw Agric Bureau, Farnham Royal.
Obaton M 1960 Les lianes ligneuses à structure anomale des forêts denses d'Afrique occidentale. *Annls Sci, Nat Sér A* **12**: 1–220
O'Dowd D J 1982 Pearl bodies as ant food: an ecological role for some leaf emergences of tropical plants. *Biotropica* **14**: 40–9
Odum E O 1953 *Fundamentals of Ecology* (1st edn). Saunders, Philadelphia
Odum H T, Pigeon R F (eds) 1970 *A Tropical Rain Forest: a Study of Irradiation and Ecology at El Verde, Puerto Rico* (3 vols). U S Atomic Energy Comm,. Washington, DC
Ofusu-Asiedu A, Cannon P 1976 *Terminalia ivorensis* decline in Ghana. *PANS* **22**: 239–42
Ogawa H *et al* 1965 Comparative ecological studies on three main types of forest vegetation in Thailand. 1 Structure and floristic composition. *Nature and Life in SE Asia* **4**: 13–48
Okafor J C 1980 Edible indigenous woody plants in the rural economy of the Nigerian forest zone. *Forest Ecol Manage* **3**: 45–55
Okali D U U 1971 Rates of dry-matter production in some tropical forest tree seedlings. *Ann Bot* **35**: 87–97
O'Keefe P 1983 Fuel for the people: fuel in the Third World. *Ambio* **12**: 115–17
Okigbo B N 1977 Farming systems and soil erosion in West Africa. In Greenland D J, Lal R (eds) *Soil Conservation and Management in the Humid Tropics*. Wiley, pp 151–62
Oldeman R A A 1974 L'architecture de la forêt guyanaise. *Mém ORSTOM* **73**, Paris
Oldeman R A A 1978 Architecture and energy exchange of dicotyledonous trees in the forest. In Tomlinson P B, Zimmermann M H (eds) *Tropical Trees as Living Systems*. Cambridge Univ Press, pp 535–60
Opeke L K 1982 *Tropical Tree Crops*. Wiley
Opler P A, Frankie G W, Baker H G 1976 Rainfall as a factor in the release, timing and synchronisation of anthesis by tropical trees and shrubs. *J Biogeog* **3**: 231–6
Oppenheimer J R, Lang G E 1969 Cebus monkeys: effects on branching of *Gustavia* trees. *Science* **165**: 187–8
OTA 1984 *Technologies to Sustain Tropical Forest Resources*. U S Congress, Office of Techn Assess **OTA-F-214**, Washington, DC
Padoch C, Vayda A P 1983 Patterns of resource use and human

settlement in tropical forests. In Golley F B (ed) *Ecosystems of the World.* **14A** *Tropical Rain Forest Ecosystems*. Elsevier, Amsterdam, pp 301–13

Percival M, Womersley J S 1975 Floristics and ecology of the mangrove vegetation of Papua New Guinea. *Bot Bull, PNG* **8**: 1–94

Peters C M 1983 Observations on Maya subsistence and the ecology of a tropical tree. *American Antiquity* **48**: 610–15

Peters C M, Pardo-Tejada E 1982 *Brosimum alicastrum* (Moraceae): Uses and potential in Mexico. *Econ Bot* **36**: 166–75

Piringer A A, Downs R J, Borthwick H A 1958 Effects of photoperiods on *Rauvolfia. Amer J Bot* **45**: 323–6

Pohl R W 1974 Blooming behavior of bamboos in Costa Rica. *Amer J Bot* **61** (5 suppl): 48–9

Poore M E D 1968 Studies in Malaysian rain forest. 1 The forest on Triassic sediments in Jengka Forest Reserve. *J Ecol* **56**: 143–96

Prance G T 1979 The taxonomy and phytogeography of the Chrysobalanaceae of the Atlantic coastal forests of Brazil. *Revta Brasil Bot* **2**: 19–39

Prance G T 1982 Forest refuges: evidence from woody Angiosperms. In Prance G T (ed) *Biological Diversification in the Tropics*. Columbia, New York, pp 137–58

Prance G T, Schubart H O R 1978 Notes on the vegetation of Amazonia. 1 A preliminary note on the origin of the open white sand *campinas* of the lower Rio Negro. **30**: 60–3

Primack R B, Tomlinson P B 1980 Variation in tropical forest breeding systems. *Biotropica* **12**: 229–31

Proctor J 1983 Mineral nutrients in tropical forests. *Progress in Physical Geog* **7**: 422–31

Pruša E 1985 Die böhmischen und mährischen Urwälder – ihre Struktur und Ökologie. Academia, Prague

Putz F E 1979 Aseasonality in Malaysian tree physiology. *Malays Forester* **42**: 1–24

Rachie K O 1983 Intercropping tree legumes with annual crops. In Huxley P A (ed) *Plant Research and Agroforestry*. Internat Council for Res in Agrofor, Nairobi, pp 103–16

Raich J W 1980 Fine roots regrow rapidly after forest felling. *Biotropica* **12**: 231–2

Rao H S 1953 Vegetative propagation and forest tree improvement. *Indian Forester* **79**: 176–83

Rath S, Das G C 1979 Effect of ringing and growth retardants on growth and flowering of mango. *Sci Hort* **10**: 101–4

Reich P B, Borchert R 1982 Phenology and ecophysiology of the tropical tree, *Tabebuia neochrysantha* (Bignoniaceae). *Ecology* **63**: 294–9

Reinders-Gouwentak C A 1965 Physiology of the cambium and other secondary meristems of the shoot. *Encycl Plant Physiol* **15**(1): 1077–1105

Rejmánek M 1976 Centres of species diversity and centres of species diversification. In Novák J, Pacltová B (eds) *Evolutionary Biology*. Academia, Prague, pp 393–410

Richards P W 1952 *The Tropical Rain Forest* (1st edn). Cambridge Univ Press

Richards P W 1983 The three-dimensional structure of tropical rain forest. In Sutton S L, Whitmore T C, Chadwick A C (eds) *Tropical Rain Forest: Ecology and Management*. Blackwell, pp 3–10

Richards P W, Tansley A G, Watt A S 1939; 1940 The recording of structure, life-form and flora of tropical forest communities as a basis for their classification. *Imp For Inst, Oxford, Paper* **19**: 1–19 (Also publ in *J Ecol* **28**: 224–39)

Risch S J, Rickson F R 1981 Mutualism in which ants must be present before plants produce food bodies. *Nature (Lond)* **291**: 149–50

Robbins R G, Wyatt-Smith J 1964 Dry land forest formations and forest types in the Malayan peninsula. *Malay Forester* **27**: 188–216

Robock A, Mass C 1982 The Mount St Helens volcanic eruption of 18 May, 1980: large surface temperature effects. *Science* **216**: 628–30

Roche L 1979 Forestry and the conservation of plants and animals in the tropics. *Forest Ecol Manage* **2**: 103–22

Roose E 1977 In Greenland DJ, Lal R (eds) *Soil conservation and management in the humid tropics*. Wiley, pp 177–87

Roth I 1984 *Stratification of Tropical Forests as Seen in Leaf Structure*. Junk, The Hague

Ruangpanit N 1971 Crown cover of hill-evergreen trees as affected to soil and water losses. *Kog-ma Watershed Res Bull, Bangkok* no. 7: 1–25

Sachs J 1887 *Lectures on the Physiology of Plants*. Oxford Univ Press

Salati E, Dall'olio A, Matsui E, Gat J R 1979 Recycling of water in the Amazon Basin: an isotopic study. *Water Resources Res* **15**: 1250–8

Salati E, Vose P B 1983 Depletion of tropical rain forests. *Ambio* **12**: 67–71

Salazar R, Rose D 1984 Firewood yields of individual trees of *Guazema ulmifolia* Lam. in pastures in Hojancha, Guanacaste

– Costa Rica. *Commonw for Rev* **63**: 271–8

Sale P J M 1970 Growth, flowering and fruiting of cacao under controlled moisture conditions. *J Hort Soc* **45**: 99–118

Sanchez P A, Bandy D E, Villachica J H, Nicholaides J J 1982 Amazon basin soils: management for continuous crop production. *Science* **216**: 821–7

Sanchez P A, Palm C A, Davey C B, Szott L T, Russell C E 1985 Tree crops as soil improvers in the humid tropics? In Cannell M G R, Jackson J E (eds) *Attributes of Trees as Crop Plants*. Inst Terrestrial Ecol, Monks Wood, Huntingdon, pp 327–58

Sasaki S 1980 Storage and germination of dipterocarp seeds. *Malays Forester* **43**: 290–308

Sasaki S, Mori T 1981 Growth responses of dipterocarp seedlings to light. *Malays Forester* **44**: 319–45

Schimper A F W 1898 *Pflanzengeographie auf physiologischer Grundlage* (1st edn). Jena. (Also 1935 3rd edn, with von Faber F C; 1903 Engl trans Fisher W R, Clarendon)

Schmitz A 1963 Aperçu sur les groupements végétaux du Katanga. *Bull Soc r Bot Belge* **96**: 233–460

Schnell R 1950 *Le forêt dense*. Lechevalier, Paris

Schnell R 1952 Contribution à une étude phytosociologique et phytogéographique de l'Afrique occidentale: les groupements et les unités géobotaniques de la région guinéenne. *Mém de l'IFAN* **18**: 41–234

Schnell R 1961 Le problème des homologies phytogéographiques entre Afrique et l'Amerique tropicales. *Mém Mus nat Hist nat, Paris, sér B* **11**(2): 137–242

Schnell R 1970; 1971 Introduction à la phytogéographie des pays tropicaux. (2 vols). Gauthier-Villars, Paris

Scholander P F, Dam L van, Scholander S I 1955 Gas exchange in roots of mangroves. *Amer J Bot* **42**: 92–8

Schulz J P 1960 *Ecological Studies on the Rain Forest in Northern Suriname*. North-Holland, Amsterdam

SCOPE 1985 *Environmental Consequences of Nuclear War. Vol II Ecological and Agricultural Effects*. Internat council of sci unions, sci cttee on problems of the env, Harwell M A, Hutchinson T C (eds). Wiley

SCOPE 1986 *Environmental Consequences of Nuclear War. Vol I Physical and Atmospheric Effects*. Pittock A B, Ackerman T P, Crutzen P J, MacCracken C, Shapiro C S, Turco R P (eds). Wiley

Sequeira L, Steeves T A 1954 Auxin inactivation and its relation to leaf-drop caused by the fungus *Omphalia flavida*. *Plant Physiol* **29**: 11–16

Shimwell D W 1971 *The Description and Classification of*

Vegetation. Sidgwick & Jackson

Shiokura T 1980 [A method to measure growth in tropical woods.] *J Agric Sci, Tokyo, Nogyo Daigaku* **25**: 27–35

Shuzeng L, Zongwei C 1981 Preliminary observations on flower bud differentiation in *Mangifera indica. Acta Hortic Sinica* **8**: 9–14

Simmonds N W 1980 Monocarpy, calendars and flowering cycles in angiosperms. *Kew Bull* **35**: 235–46

Simon S V 1914 Studien über die Periodizität der Lebensprozesse der in dauernd fechten Tropengebieten heimischen Bäume. *Jahrb wiss Bot* **54**: 71–187

Sinclair R 1984 Water relations of tropical epiphytes. III Evidence for Crassulacean acid metabolism. *J Exp Bot* **35**: 1–7

Singh A K, Singh M K, Mascarenhas O A J 1985 Community forestry for revitalising rural ecosystems: a case study. *Forest Ecol Manage* **10**: 209–32

Singh K, Nanda K K 1981 Effect of photoperiod on extension growth, rest period and flowering of *Bauhinea acuminata* L. seedlings. I Analysis of the first year growth. *Indian J Exp Biol* **19**: 337–40

Sivard R L 1983 *World Military and Social Expenditures 1983*. World Priorities, Leesburg, Virginia

Sleigh P A, Collin H A, Hardwick K 1984 Distribution of assimilate during the flush cycle of growth in *Theobroma cacao* L. *Plant Growth Reg* **2**: 381–91

Smith A P 1972 Buttressing of tropical trees: a descriptive model and new hypotheses. *Amer Nat* **106**: 32–46

Smith A P 1974 Bud temperature in relation to nyctinastic leaf movement in an Andean giant rosette plant. *Biotropica* **6**: 263–6

Smith J A C, Griffiths H, Bassett M, Griffiths N M 1985 Day-night changes in the leaf water relations of epiphytic bromeliads in the rain forests of Trinadad. *Oecologia* **67**: 475–85

Soepadmo E, Eow, B K 1976 The reproductive biology of *Durio zibethinus* Murr. *Gardens Bull, Singapore* **29**: 25–33

Sollins P, Drewry G 1970 Electrical conductivity and flow rate of water through the forest canopy. In Odum H T, Pigeon R F (eds) *A Tropical Rain Forest: a Study of Irradiation and Ecology at El Verde, Puerto Rico* (Book 3). US Atomic Energy Comm, pp F29–33

Sommer A 1976 Attempt at an assessment of the world's tropical forests. *Unasylva* **28**: 5–24

Stark N M, Jordan C F 1978 Nutrient retention by the root mat of an Amazonian rain forest. *Ecology* **59**: 434–7

Steingraeber D A, Fisher J B 1986 Indeterminate growth of leaves in *Guarea* (Meliaceace): a twig analogue. *Amer. J. Bot.* **73**: 852–62

Stubblebine W, Langenheim J H, Lincoln D 1978 Vegetative response to photoperiod in the leguminous tree *Hymenaea courbaril* L. *Biotropica* **10**: 18–29

Sutton S L 1983 The spatial distribution of flying insects in tropical rain forests. In Sutton S L, Whitmore T C, Chadwick, A C *Tropical Rain Forest: Ecology and Management*. Blackwell, pp 77–91

Sutton S L, Whitmore T C, Chadwick A C 1983 *Tropical Rain Forest: Ecology and Management*. Blackwell

Swaine M D, Beer T 1977 Explosive seed dispersal in *Hura crepitans* (Euphorbiaceae). *New Phytol* **78**: 695–708

Swaine M D, Hall J B 1983 Early succession on cleared forest land in Ghana. *J Ecol* **71**: 601–27

Swaine M D, Hall J B, Lieberman D 1984 Leaf dynamics in tropical forest trees. In Chadwick A C, Sutton S L (eds) *Tropical Rain Forests – the Leeds Symposium*. Leeds Phil & Lit Soc, Abstr 68, pp 321–2

Synge H (ed) 1981 *The Biological Aspects of Rare Plant Conservation*. Wiley

Tanner E V J 1982 Species diversity and reproductive mechanisms in Jamaican trees. *Biol J Linn Soc* **18**: 263–78

Tansley A G 1935 The use and abuse of vegetational concepts and terms. *Ecology* **16**: 284–307

Taylor C E 1982 Reproductive biology and ecology of some tropical pioneer trees. PhD Thesis, Aberdeen Univ

Taylor C J 1960 *Synecology and Silviculture in Ghana*. Nelson

Taylor C J 1962 *Tropical Forestry*. Oxford Univ Press

Thornton I W B 1984 Krakatau – the development and repair of a tropical ecosystem. *Ambio* **13**: 216–25

Tingwa P O, Young R E 1975 The effect of indole-3-acetic acid and other growth regulators on the ripening of avocado fruits. *Plant Physiol* **55**: 937–40

Tixier P 1966 *Flore et végétation orophiles de l'Asie tropicale*. Soc d'édition d'enseignement supérieur, Paris

Tomlinson P B 1978 Branching and axis differentiation in tropical trees. In Tomlinson P B, Zimmermann M H (eds) *Tropical Trees as Living Systems*. Cambridge Univ Press, pp 187–207

Tomlinson P B 1980 *The Biology of Trees Native to Tropical Florida*. Harvard Forest, Petersham, Mass

Tompsett P B 1982 The effect of desiccation on the longevity of seeds of *Araucaria hunsteinii* and *A. cunninghamii*. *Ann Bot.* **50**: 693–704

UNESCO 1978 *Tropical Forest Ecosystems*. UNESCO/UNEP/

FAO. UNESCO, Paris, p 286

USDA 1967 *Supplement to Soil Classification System, 7th Approximation*. Soil Cons Service, US Dept Agric, Washington, DC

van Steenis C G G J 1958 Basic principles of rain forest sociology. In *Study of Tropical Vegetation*. UNESCO, Paris, pp 159–65

Vázquez-Yanes C 1974 Studies on the germination of seeds of *Ochroma lagopus, Turrialba* **24**: 176–9

Vázquez-Yanes C 1977 Germination of a pioneer tree (*Trema guineensis* Ficalho) from equatorial Africa. *Turrialba* **27**: 301–2

Vázquez-Yanes C, Orozco Segovia A 1984 Ecophysiology of seed germination in the tropical humid forests of the world: a review. In Medina E, Mooney H, Vázquez-Yanes C (eds) *Physiological Ecology of Plants of the Wet Tropics*. Junk, The Hague

Vázquez-Yanes C, Smith H 1982 Phytochrome control of seed germination in the tropical rain forest pioneer trees *Cecropia obtusifolia* and *Piper auritum* and its ecological significance. *New Phytol* **92**: 477–85

Vogel M 1975 Recherche du determinisme du rhythme de croissance du cacaoyer. *Café, Cacao, Thé* **19**: 265–90

Volatron B 1976 La mise en valeur des richesses forestières en Amazonie Bresilienne et en Colombie. *Bois et Fôrets Tropiques* **165**: 59–76; **166**: 55–70

Wadsworth F H 1983 Production of usable wood from tropical forests. In Golley F B (ed) *Tropical Rain Forest Ecosystems*. Elsevier, Amsterdam, pp 279–88

Walter H 1979 *Vegetation of the Earth and Ecological Systems of the Geobiosphere* (2nd edn, transl Wieser J). Springer, New York

Walter H, Lieth H 1960–67 *Klimadiagramm-Weltatlas*. Fischer, Jena

Wareing P F 1969 The control of bud dormancy in seed plants. *Symp Soc Exptl Biol* **23**: 241–62

Warming J E B 1895 *Plantesamfund, Grundtraek af den Økologiske Plantegeografi*. Philipsen, Copenhagen (Also Engl trans 1909 *Oecology of Plants: an Introduction to the Study of Plant Communities*. Clarendon)

Watanabe S, Sibaoka T 1983 Light-induced and auxin-induced leaflet opening in detached pinnae of *Mimosa pudica. Plant and Cell Physiol* **24**: 641–8

Waterman P G 1983 Distribution of secondary metabolites in rain forest plants: towards an understanding of cause and effect. In Sutton S L, Whitmore T C, Chadwick A C (eds) *Tropical Rain Forest: Ecology and Management*. Blackwell, pp 167–79

Webb L J, Tracey J G, Haydock K P 1967 A factor toxic to seedlings of the same species associated with living roots of the non-gregarious subtropical rain forest tree. *Grevillea robusta. J Appl Ecol* **4**: 13–25

Went F W 1957 *The Experimental Control of Plant Growth*. Ronald Press, New York

Went F W, Stark N M 1968 Mycorrhiza. *BioScience* **18**: 1035–9

Whitmore J L, Hinojosa G 1977 Mahogany (*Swietenia*) hybrids. *Forest Serv Res Pap* **ITF-23**, Inst Trop Forestry, Rio Piedras, Puerto Rico

Whitmore T C 1962a Studies in systematic bark morphology. I Bark morphology in the Dipterocarpaceae. *New Phytol* **61**: 191–207

Whitmore T C 1962b II General features of bark construction in Dipterocarpaceae. *New Phytol* **61**: 208–20

Whitmore T C 1975 *Tropical Rain Forests of the Far East* (1st edn). Oxford Univ Press

Whitmore T C 1983 Secondary succession from seed in tropical rain forests. *Forestry Abstr* **44**: 767–79

Wiebes J T 1979 Co-evolution of figs and their insect pollinators. *Ann Rev Ecol Syst* **10**: 1–12

Wilkinson G 1939 Root competition and silviculture. *Malay Forester* **8**: 11-15

Williamson G B *et al* 1983 Drip-tips, drop size and leaf drying. *Biotropica* **15**: 232–4

Withers L A 1980 *Tissue Culture Storage for Genetic Conservation*. Internat Bd for Plant Genet Resources, FAO, Rome

Witkamp M 1970 Aspects of soil microflora in a gamma-irradiated rain forest. In Odum H T, Pigeon R F (eds) *A Tropical Rain Forest: a Study of Irradiation and Ecology at El Verde, Puerto Rico* (Book 2). U S Atomic Energy Comm, pp F29–33

Wood P J 1979 Variation in tropical tree species. In Ford E D, Malcolm D C, Atterson J (eds) *The Ecology of Even-Aged Plantations*. Institute of Terrestrial Ecol, Cambridge, pp 567–76

Woodwell G et al 1983 Global deforestation: contribution to atmospheric carbon dioxide. *Science* **222**: 1081–6

Wyatt-Smith J 1953 A note on the vegetation of some islands in the Malacca Straits. *Malay Forester* **16**: 191–205

Wyatt-Smith J 1966 Ecological studies on Malayan forests I. *Malay Forestry Dept, Res Pamphl* **32**

Wycherley P R 1973 The phenology of plants in the humid tropics. *Micronesica* **9**: 75–96

Yap S K 1982 The phenology of some fruit tree species in a lowland dipterocarp forest. *Malays Forester* **45**: 21–32

Yoda K 1974 Three dimensional distribution of light intensity in a tropical rain forest of W Malaysia. *Jap J Ecol* **24**: 247–54
Zahl P A 1964 Malaysia's giant flowers and insect-trapping plants. *Nat Geographic Mag* **125**: 681–97
Zimmermann M H, Wardrop A B, Tomlinson P B 1968 Tension wood in the aerial roots of *Ficus benjamina* L. *Wood Sci and Technol* **2**: 95–104

Species index

Numbers in **bold** type refer to Figures and Tables. *See also* Table 4.7, p. 118 for representative woody plant genera in different tropical forest regions.

General index

Numbers in **bold** type refer to Figures and Tables.